Pr

One of the most im
of Christian teachin
human soul. J. P. Moreland provides a robust philosophical and theological defense of the soul, but just as importantly, demonstrates why understanding the nature of soul is *critical* to many of the cultural discussions of today. Moreland offers us a rare combination: philosophical acuity, philosophical fidelity, and technical accessibility. I would highly recommend this to anyone!

—**Mike Erre**
Lead Pastor of EvFree Fullerton
Author of *Jesus of Suburbia* and *Astonished*

Many think science has displaced the soul. Nothing could be further from the truth! In this quintessentially Moreland book, *The Soul* provides compelling evidence for the existence of the soul and shows why it matters so deeply for our daily lives. Moreland's teaching on the soul has transformed my personal and academic life. I happily recommend that it be read and studied by pastors, educators, as well as apologists and theologians alike.

—**Sean McDowell**
Assistant Professor, Biola University
Coauthor of *Is God Just a Human Invention?*

Increasingly, secular thought denies the human soul. In this splendid book, J. P. Moreland restores the soul to the center of our identity as creatures made in the image of God. Moreland is a first-rate philosopher of mind and top-tier Christian apologist, known for his rigorous argumentation, but here the style is straightforward and accessible to non-specialists. All of the key terms are clearly defined, and then the biblical, theological, and philosophical case for the soul is laid out. Anyone who suspects the soul has been dismissed without fair trial will be encouraged by Moreland's bold and timely defense.

—**Angus Menuge**
President of the Evangelical Philosophical Society (www.epsociety.org)

THE SOUL

HOW WE KNOW IT'S REAL AND WHY IT MATTERS

J. P. MORELAND

MOODY PUBLISHERS
CHICAGO

Edited by Christopher Reese
Interior design: Ragont Design
Cover design: Erik M. Peterson

Library of Congress Cataloging-in-Publication Data

Moreland, James Porter, 1948-
The soul : how we know it's real and why it matters / J.P. Moreland.
pages cm
Includes bibliographical references.
ISBN 978-0-8024-1100-6
1. Soul--Christianity. 2. Spirit. 3. Consciousness--Religious aspects--Christianity. 4. Theological anthropology--Christianity. 5. Religion and science. I. Title.
BT741.3.M67 2014
233'.5--dc23
2013045931

We hope you enjoy this book from Moody Publishers. Our goal is to provide high-quality, thought-provoking books and products that connect truth to your real needs and challenges. For more information on other books and products written and produced from a biblical perspective, go to www.moodypublishers.com or write to:

Moody Publishers
820 N. LaSalle Boulevard
Chicago, IL 60610

1 3 5 7 9 10 8 6 4 2

Printed in the United States of America

To Dallas Willard,

a man with the largest soul I ever encountered.

CONTENTS

Introduction: WHAT'S SO IMPORTANT ABOUT THE SOUL AND CONSCIOUSNESS?

Throughout history, the vast majority of people, educated and uneducated alike, have been dualists (those who believe that the soul is an immaterial thing different from the body and brain), at least in the sense that they have taken a human to be the sort of thing that could enter life after death while his or her corpse was left behind. Some form of *dualism appears to be the natural response to what we seem to know about ourselves through introspection and in other ways (words preceded by an asterisk are defined at the end of the chapter, and in the glossary at the end of the book). Many thinkers who deny dualism still admit that it is the commonsense view. Thus, physicalist Jaegwon Kim (a physicalist is someone who denies the existence of a soul and says that consciousness is merely physical or at least dependent on the physical) acknowledges that "We commonly think that we, as persons, have a mental and bodily dimension. . . . Something like this dualism of personhood,

I believe, is common lore shared across most cultures and religious traditions . . ."[1]

Not only has belief in the soul been included in the commonsense beliefs of people all over the world throughout the ages, but also the idea that there is a soul has been the constant teaching of the Christian church since its beginning. Throughout church history, Christianity has given affirmative answers to questions about the reality of the three great topics of Western philosophy—namely, God, the soul, and life everlasting. For centuries, most Christian thinkers have believed in the souls of men and beasts, as it used to be put. Animals and humans are (or have) an immaterial entity—a soul, a life principle, a ground of sentience—and they have a body. More specifically, a human person is a functioning unity of two distinct entities, body and soul. The human soul, while not by nature immortal, is nevertheless capable of entering a disembodied intermediate state upon death, however incomplete and unnatural this state may be, and, eventually, being reunited with a resurrected body.

Today, however, it is widely believed that science has rendered this commonsense and biblical view obsolete and implausible. As Christian physicalist Nancey Murphy says, even though science cannot prove dualism is false, still, "science has provided a massive amount of evidence suggesting that we need not postulate the existence of an entity such as a soul or mind in order to explain life and consciousness."[2] Sadly, this opinion is not limited to academic circles. To see this, consider the following two cases.

Case 1: A few years ago, *Time* magazine featured an article defending stem-cell research on human embryos:

"These [embryos] are microscopic groupings of a few differentiated cells. There is nothing human about them, except potential—and, if you choose to believe it, a soul."[3] This expresses a widely held opinion that when it comes to belief in the soul, you're on your own. There is no evidence one way or another. You must choose arbitrarily or, perhaps, on the basis of private feelings. For many, belief in the soul is like belief in ghosts—an issue best left to the pages of the *National Enquirer*.

Case 2: *The Walking Dead* is a very popular television show today. In the first season's final episode, a scientist shows a group of ordinary people a video of the inside workings of a live human brain. It looks like a complex web of wires and nodes, with a multitude of flashing lights traveling to and fro. He then declares matter-of-factly that all of the electrical activity that they see is actually the real you. When those "lights" go off, you cease to exist.[4]

Regardless of how often this mantra is recited, nothing could be further from the truth. In reality, a robust case can be offered for the view that consciousness and the soul are *immaterial*—not physical—realities. Thinking through these issues is a fascinating adventure of considerable importance. French philosopher Blaise Pascal rightly remarked that the soul's nature is so important that one must have lost all feeling not to care about the issue.[5] The great Presbyterian scholar J. Gresham Machen once observed, "I think we ought to hold not only that man has a soul, but that it is important that he should know that he has a soul."[6] But why should we think Machen is correct about this? Why is it worth spending time learning about the immaterial nature of consciousness and the soul? After all, life is busy and we have many demands on our time. I believe there are at least four reasons why this topic is worthy of our attention.

First, the Bible seems to teach that consciousness and the soul are immaterial and we need to regard this teaching as genuine knowledge and not as faith commitments that we merely hope are true. For twenty centuries, the vast majority of educated and uneducated Christians understood the Bible in this way. The historic Christian position is nicely stated by H. D. Lewis: "Throughout the centuries Christians have believed that each human person consists in a soul and body; that the soul survived the death of the body; and that its future life will be immortal."[7] (For a detailed discussion of the biblical evidence, see chapter 2.) Lay people are rightly suspicious of and concerned about what they see as politically correct revisions of what the church has held for centuries, even if the revisions are done in the name of science.

It is very important for Christians to understand the central teachings of Scripture to be sources of **knowledge* and not merely truths to be accepted by a blind act of faith. Why? It is on the basis of knowledge (or perceived knowledge)—not faith, commitment, or sincerity—that people are given the right to lead, act in public, and accomplish important tasks. We give certain people the right to fix our cars, pull our teeth, write our contracts, and so on because we take those people to be in possession of the relevant body of knowledge. Moreover, it is the possession of knowledge, and not mere truth alone, that gives people confidence and courage to lead, act, and risk. Accordingly, it is of crucial importance that we promote the central teachings of Christianity in general, and the spiritual nature of consciousness and the soul in particular, as a body of knowledge and not as a set of faith-practices to be accepted on the basis of mere belief or simply a shared narrative. To fail at this point is to risk being marginalized and disregarded as those promoting a privatized set of feelings or desires that fall short of knowledge.

Unfortunately, the contemporary cultural milieu—inside and outside the church—in which we live and move and have our being is precisely one that takes the central teachings of the Bible to be on the order of astrology or the Flat Earth Society. If we revise the traditionally understood teachings of Scripture in light of the supposed demands of science, then we contribute to the idea that it is really science that gives us confident knowledge of reality and not Scripture. Such an attitude will undermine our efforts to reach the lost and pass on our Christian faith to our children.

Here's a simple definition of knowledge: *To represent reality in thought or experience the way it really is on the basis of adequate grounds.* To know something (the nature of God, forgiveness, cancer) is to think of or experience it as it really is on a solid basis of evidence, experience, intuition, and so forth. Little can be said in general about what counts as "adequate grounds." The best one can do is to start with specific cases of knowledge and its absence in art, chemistry, memory, Scripture, logic, introspection, etc., and formulate helpful descriptions of "adequate grounds" accordingly.

Please note that *knowledge has nothing to do with certainty or an anxious quest for it.* One can know something without being certain about it, and in the presence of doubt or with the admission that one might be wrong. Recently, I know that God spoke to me about a specific matter but I admit it is possible I am wrong about this (though, so far, I have no good reason to think I am wrong). When Paul says, "This you know with certainty" (Eph. 5:5), he clearly implies that one can know without certainty; otherwise, the statement would be redundant. How so? If I say, "Give me a burger with pickles on it," I imply that it is possible to have a burger without pickles. If, contrary to fact, pickles were simply essential ingredients of burgers, it would be redundant to ask for burgers with pickles. The parallel to "knowledge with certainty" should be easy to see. When Christians claim to have knowledge of this or that—for example, that God is real, that Jesus rose from the dead, that the Bible is the Word of God—they are not saying that there is no possibility that they could be wrong, that they have no doubts, or that they have answers to every question raised against them. They are simply saying that these and other claims satisfy the definition given above (that is, to represent reality in thought or experience the way it really is on the basis of adequate grounds).

The deepest issue facing the church today is this: Are its main creeds and central teachings items of knowledge, or mere matters of blind faith, privatized personal beliefs, or issues of feeling to be accepted or set aside according to the individual or cultural pressures that come and go? Do these teachings have cognitive and behavioral authority that set a worldview framework for approaching science, art, or ethics—indeed, all of life? Or is cognitive and behavioral authority set by what scientists or the American Psychiatric Association say, or by what Gallup polls tell us is embraced by cultural and intellectual elites? Do we turn to these sources and then set aside or revise two thousand years of Christian thinking and doctrinal/creedal expressions in order to make Christian teaching acceptable to the neuroscience department at UCLA? The question of whether or not Christianity provides its followers with a range of knowledge is no small matter. It is a question of authority for life and death, and lay brothers and sisters are keeping a watchful eye on Christian thinkers and leaders to see how they approach this matter. Currently, the nature of consciousness and the soul are at the heart of the struggle for the intellectual authority of extra-scientific fields such as theology and the Bible.

Second, as the Time *article cited above implies, the reality of the soul is important to various ethical issues that crucially involve an understanding of human persons.* There is a deep connection between the reality of the human soul and the sort of high, intrinsic value human persons possess, and this is relevant to ethical reflection—for example, in areas such as abortion, euthanasia, and human rights. Now, one could believe in the soul and reject this sort of value, and one could reject the soul and embrace this sort of value. Whether or not each view can be justified is, of course, another matter. But, in any case, the existence of the soul factors into a good deal of bioethical argumentation. It is in virtue of the type of soul humans have, reflecting as it does the image of God, that humans have such high, intrinsic value. As I see it, physicalists of various stripes have a difficult task in attempting to justify this sort of value for human persons insofar as they are material objects. If it is true that we are merely physical objects, we are of little value, or so it seems to me.[8]

But, you may respond, what of the body? Are you being

Platonistic or gnostic here and devaluing the body's worth? No, not at all. I just don't think the body is of much value simply insofar as it is physical. As I see it, the body has value for these reasons: (1) I take the body to be an ensouled, spatially extended, physical structure; thus, the body includes the soul to be a body, and it is of value accordingly. A body without a soul in it is just a corpse. By contrast with a body, a corpse is of little intrinsic value. (2) It has certain qualities (color, smell, sound, taste, texture) that serve as the *metaphysical grounding of many of its aesthetic properties, and neither these qualities nor aesthetic properties are physical (that is, *properties like colors and beauty and others mentioned below are *abstract). A major reason bodies are beautiful is due to these kinds of qualities. (3) It possesses certain geometrical features, for example, shape and symmetry, but these are not physical, they are abstract. (4) It possesses a certain complexity of arrangement, but since this complexity is an abstract object, it is not physical. (5) It is owned by the person, it is the vehicle in virtue of which the person is known, and it is intimately and causally related to the person. (6) It exists and, insofar as any existing thing has value, it does. (7) It is physical.

(6) and (7) involve little value compared to the other factors. It must be kept in mind that the intrinsic value/beauty of the creation is due to factors like (2)–(4) above, not (7). Further, (1) and (5) are the key reasons why the body has value as is evident when we compare the value of the body as an ensouled structure that is intimately related to the soul to the value of a corpse or the body of a mindless zombie for which these value-making features are absent.

Third, loss of belief in life after death is related to a commitment to the authority of science above theology, along with a conviction that belief in the soul is scientifically discredited. As philosopher John Hick pointed out, "This considerable decline within society as a whole, accompanied by a lesser decline within the churches, of the belief in personal immortality clearly reflects the assumption within our culture that we should only believe in what we experience, plus what the accredited sciences certify to us."[9] While there are, indeed, a small group of Christian physicalists who try to make sense of of life after death without a soul, most people rightly understand the afterlife as involving the departure of the soul at death (and Christianity, of course, teaches that after this disembodied intermediate state we will all receive resurrection bodies).[10] For most people it just stands to reason that if there is no soul, as modern science would have us believe, there is no life after death.

Finally, in The Divine Conspiracy, *Dallas Willard says that understanding the immaterial nature of the human spirit is crucial to grasping the essence of spiritual growth*: "To understand spirit as 'substance' is of the utmost importance in our current world, which is so largely devoted to the ultimacy of matter."[11] Willard argues that without a careful grasp of the soul's nature, it becomes virtually impossible to develop a detailed model of spiritual formation. Thus, knowledge of the existence and nature of the soul is crucial for our self-understanding and for developing a view about how to mature in the Way of Jesus.

For these four reasons, it is crucial that parents, pastors, and parishioners recapture the confidence that in

Christianity we are presented with genuine knowledge about the existence and nature of the soul. It is also important for us to regain knowledge of how to make a case for the immaterial nature of consciousness and the soul without using the Bible. While in past decades Scripture was held to be a source of authority in our culture, it no longer is today. If we are to make a persuasive case for Christianity, and aspects of its teaching like the soul, we must learn to use reason and evidence to defend biblical doctrine. In the chapters to follow, we will be concerned with making a reasonable case for an immaterial soul and consciousness. But first we need to get before us a set of distinctions that will help in the discussion to follow, and we will examine these in chapter 1.

CHAPTER IN REVIEW

In this chapter we saw that we should care about the soul and consciousness because:

- The Bible seems to teach that consciousness and the soul are immaterial and we need to regard this teaching as genuine.
- The reality of the soul is important to various ethical issues.
- Loss of belief in life after death is related to a commitment to the authority of science above theology, and the conviction that belief in the soul is scientifically discredited.
- Understanding the immaterial nature of the human spirit is crucial to grasping the essence of spiritual growth.

KEY VOCABULARY

Abstract: In the discipline of philosophy, this term refers to properties (e.g., redness, hardness; see also the entry for "property" below) and relations (e.g., taller than, heavier than) that do not exist by themselves in space or time, but can exist potentially *in* many different places and times (e.g., *redness* can exist *in* both an apple and a ball; a dog can be *heavier than* a rock, while a rock can be *heavier than* a leaf). This term is often contrasted with spatial/temporal objects that are *concrete* or *physical* (e.g., a house, a cow, a bracelet).

Dualism: The view that the soul is an immaterial thing different from the body and brain.

Knowledge: To represent reality in thought or experience the way it really is on the basis of adequate grounds.

Metaphysics: In philosophy, this refers to the study of the most fundamental aspects of reality that underlie what we experience through our senses. Common topics of study in metaphysics include existence, substance, properties, causation, events, and mind/body questions.

Physicalism: The view that the only things that exist are physical substances, properties, and events. In relation to humans, the physical substance is the material body, especially the brain and central nervous system.

Property: an existent reality that is universal, immutable, and can (or perhaps must) be *in* or *had* by other things more basic, such as a substance. Thus, a cow (substance) can have the property of being brown. The brownness (property) is had by the cow (the substance).

NOTES

1. Jaegwon Kim, "Lonely Souls: Causality and Substance Dualism," in *Soul, Body and Survival*, ed. Kevin Corcoran (Ithaca, NY: Cornell University Press, 2001), 30.
2. Nancey Murphy, "Human Nature: Historical, Scientific, and Religious Issues," in *Whatever Happened to the Soul?* eds. Warren S. Brown, Nancey Murphy, and H. Newton Malony (Minneapolis: Fortress Press, 1998), 18.
3. Michael Kinsley, "If You Believe Embryos Are Humans . . ." *Time*, June 25, 2001, 80.
4. Adam Fierro and Frank Darabont, "TS-19," *The Walking Dead*, season 1, episode 6, directed by Guy Ferland, aired December 5, 2010 (New York: AMC, 2010), Netflix. I want to thank Michael Sanborn for pointing this out to me.
5. Blaise Pascal, *Pensees*, section III, 194.
6. J. Gresham Machen, *The Christian View of Man* (New York: Macmillan, 1937), 159.
7. H. D. Lewis, *Christian Theism* (Edinburgh: T & T Clark, 1984), 125.
8. For a Christian physicalist defense of human value and dignity, see Kevin Corcoran, *Rethinking Human Nature* (Grand Rapids, MI: Baker, 2006), chapter 4.
9. John H. Hick, *Death & Eternal Life* (San Francisco: Harper & Row, 1980), 92.
10. For an exposition and critique of physicalist attempts to make sense of an afterlife, see William Hasker, "Materialism and the Resurrection," *European Journal for Philosophy of Religion* 3 (2011): 83–103; Jonathan Loose, "Constitution and the Falling Elevator: The Continuing Incompatibility of Materialism and Resurrection Belief," *Philosophia Christi* 14 No. 2 (2012): 439–49.
11. Dallas Willard, *The Divine Conspiracy* (San Francisco: Harper, 1998), 82.

Chapter One: A TOOLBOX FOR THE SOUL

Currently, there are two main positions taken on the *mind/body problem, as illustrated in the chart below.

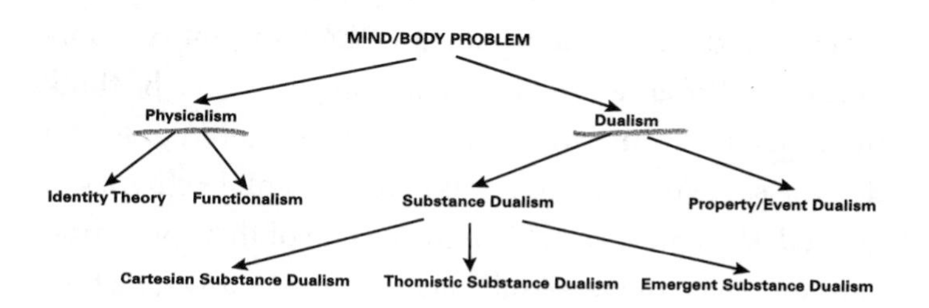

The details of the chart are not important for now, and we will unpack them in later chapters. For present purposes, note that the two main views are *physicalism and dualism. The former claims that a human being is completely physical, whereas the latter maintains that a human being is, in some sense or other, both physical and mental. Dualism comes in two major varieties: *substance dualism and *property/event dualism (more on this later). Physicalism comes in different varieties as well, but we will not explore them here. Our present purpose is to examine three key concepts that are essential to understanding the mind/body debate, and then briefly contrast dualism and physicalism. I will begin by clarifying the nature of *substances, properties, and *events.

SUBSTANCES

A *substance* is an entity like an acorn, an electron, a dog, or an angel. A human person is a substance. Substances have a number of important characteristics. First, substances are *particular, individual things.* A substance, like a particular acorn, cannot be in more than one place at the same time.

Second, a substance is a *continuant*—it can change by gaining new properties and losing old ones, yet it remains the same thing throughout the change. An acorn can go from green to red, yet the acorn itself is the same entity before, during, and after the change. A human person can be thinking about lunch and later thinking about something else, but it is the same person engaging in both mental activities. In general, substances can change in some of their properties and yet remain the same substance. That very acorn that was green is the same acorn that is now red.

Third, substances are *basic, fundamental existents.* They are not *in* other things or had by other things. My dog Fido is not in or had by something more basic than he. Rather, properties (and parts) are *in* substances that have them. For example, Fido has the property of brownness and the property of weighing twenty-five pounds. These properties are in the substance called Fido.

Fourth, substances are *unities of parts, properties, and capacities* (dispositions, tendencies, potentialities). Fido has a number of properties like the ones already listed. He also has a number of parts—four legs, some teeth, two eyes.

Further, he has some capacities or potentialities that are not always actual. For example, he has the capacity to bark even when he is silent. As a substance, Fido is a unity of all the properties, parts, and capacities had by him.

Finally, a substance has *causal powers*. It can do things in the world. A dog can bark, an acorn can hit the ground. Substances can cause things to happen.

PROPERTIES

In addition to substances, there are also entities that exist called *properties.* A property is an existent reality, examples of which are brownness, triangularity, hardness, wisdom, painfulness, being a neuron. As with substances, properties have a number of important features.

One feature is that a property is a *universal* that can be in more than one thing at the same time. Redness can be in a flag, a coat, and an apple all at once. The very same redness can be the color of several particular things all at the same time. Or, to take another example, roundness can simultaneously be in a watch, a wheel, and a pizza.

Another feature of properties is their immutability. When a leaf goes from green to red, the *leaf* changes by losing an old property and gaining a new one. But the property of redness does not change and become the property of greenness. Properties can come and go, but they do not change in their internal constitution or nature.

Moreover, properties *can, or perhaps must, be in or had by other things more basic than they.* Properties are in the things that have them. For example, redness is in the apple. The apple has the redness. One does not find redness existing all by itself. In general, when we are talking about a property, it makes sense to ask the question, "What is it that has that property?" That question is not appropriate for substances, for they are among the things that have the properties. Substances have properties; properties are had by substances.

EVENTS

Finally, there are entities in the world called *events.* Events are temporal states that occur in the world. Examples of events are a flash of lightning, the dropping of a ball, the having of a thought, the firing of a neuron, the change of a leaf, and the continued possession of sweetness by an apple (this would be a series of events). Events are temporal states or changes of states of substances. An event is the coming or going of a property in a substance at a particular time, or the continued possession of a property by a substance throughout a time. "This shirt being green now" and "this acorn changing shape then" are both examples of events. The central identifying feature of an event is the property involved in that event. For example, the event of "this shirt being green now" crucially involves the property of being green. Any event that failed to involve that property could not be the event of "this shirt being green now."

PHYSICALISM VS. DUALISM

Physicalism

Keeping these critical distinctions in mind, we can now move on to consider in more detail the basic mind/body views listed in our chart. Let's look at physicalism first.

According to strict physicalism, a human being is merely a physical entity.[1] The only things that exist are physical substances, properties, and events. When it comes to humans, the physical substance is the material body, especially the parts called the brain and central nervous system. The physical substance called the brain has physical properties, such as a certain weight, volume, size, electrical activity, chemical composition, and so forth.

There are also physical events that occur in the brain. For example, the brain contains a number of elongated cells that carry various impulses. These cells are called neurons. Various neurons make contact with other neurons through connections or points of contact called synapses. C-fibers are certain types of neurons that innervate the skin (supply the skin with nerves) and carry electrical impulses associated with pain. So when someone has an occasion of pain or an occurrence of a thought, physicalists hold that these are merely particular physical events—events where certain C-fibers are firing or certain electrical and chemical events are happening in the brain and central nervous system.

Thus, physicalists believe that we are merely a physical substance (a brain and central nervous system, a body) that has physical properties and in which occur physical events. My conscious mental life of thoughts, emotions, and pains is nothing but a stream of physical events in my brain and nervous system. The neurophysiologist can, in principle, describe these events solely in terms of C-fibers, neurons, and the chemical and physical properties of the brain. For the physicalist, I am merely a functioning brain and central nervous system enclosed in a physical body. I am a material substance characterized completely by physical properties and in which occur merely physical events, a creature made of matter—nothing more, nothing less.

What is matter? we might ask. There is no clear definition of matter, but examples of it are not hard to come by. Material objects are things like computers, carbon atoms, brains, and billiard balls. Material properties are things like negative charge, mass, and extension. Material events are items like the occurrence of a flash of lightning, the moving of an electron, the firing of a neuron in the brain.

To say more about material (or physical) properties, they are (1) publicly accessible in the sense that no one person is better suited to have private access to a material property than anyone else; any way you have available to you to know about the presence or nature of a material property (say, the weight of a chair), I have available to me as well; (2) such that an object must be either spatially located or extended to have a material property; (3) such that when a strictly material object has physical properties, that object does not engage in genuinely teleological behavior—that is, it does not undergo change for the sake of some end, purpose, goal, or final cause. Physical properties are the properties that one finds listed in chemistry or physics books. They are properties such as hardness; occupying and moving through space; having a certain shape; possessing certain chemical, electrical, magnetic, and gravitational properties; having density and mass; and being breakable, malleable, and elastic. A physical event would be the possession, coming, or going of one or more of these properties by a physical substance (or among physical substances).

Another very crucial observation to make about material substances, properties, and events is this: *No material thing presupposes or requires reference to consciousness for it to exist or be characterized.* You will search in vain through a physics or chemistry textbook to find consciousness included in any description of matter. A completely physical description of the world would be in the third person and would not include any terms that make reference to or characterize the existence and nature of consciousness. Assume that matter is actually like what our chemistry and physics books tell us it is. Now imagine that there is no God and picture a universe in which no conscious, living beings had emerged. In such an imaginary world, there would be no consciousness anywhere in the universe—no selves, sensations, beliefs, or thoughts. However, in this imaginary world, matter would still exist and be what scientists tell us it is. Carbon atoms would still be carbon atoms, electrons would still have negative charge. An electron is still an electron regardless of whether or not conscious minds exist in the world. In such a world, there could be mindless zombies with brains and nervous systems but without consciousness. This is what we mean when we say that the existence and nature of matter are independent of the existence of consciousness.

Dualism

Dualists disagree with physicalists. According to them, genuinely mental entities are real. As with matter, it is hard to give a *definition* of mental entities to which all philosophers and scientists would agree. The most popular definition of a mental property or event is one in which the subject who is having it has privileged access, that is, a way of knowing it (through introspectively experiencing it in the first person) that is not available to anyone else (someone else cannot know directly by introspection what my mental states are). Physical properties like being square or hard and physical events like a flash of lightning are such that no one person has a special way of knowing something about it. Whatever ways you have for knowing something about a flash of lightning (measuring it, taking a picture of it) are available to me and vice versa.

While there is some dispute about a definition of the mental, *examples* of mental entities are easy to supply. First, there are various kinds of *sensations:* experiences of colors, sounds, smells, tastes, textures, pains, and itches. Sensations are individual things that occur at particular times. I can have a sensation of red after looking in a certain direction or by closing my eyes and daydreaming. An experience of pain will arise at a certain time, say, after I am stuck with a pin.

Further, sensations are natural kinds of things that have, as their very essence, the felt quality or sensory property that makes them what they are. Part of the very essence of a pain is the felt quality it has, which is very different from an itch or a taste; part of the very essence of a red sensation is the presentation of a particular shade of color to my consciousness, which is quite different from a smell. Sensations are not identical to things outside a person's body—for instance, a feeling of pain is not the same thing as being stuck with a pin and shouting, "Ouch!" Sensations are essentially characterized by a certain conscious feel, and thus, they presuppose consciousness for their existence and description. If there were no conscious beings, there would be no sensations.

A second type of mental entity is called a **propositional attitude:* having a certain mental attitude involving a *proposition that is part of a "that-clause." For example, one can hope, desire, fear, dread, wish, think, or believe that *P*, where *P* may be the proposition, "The Royals are a great baseball team." A proposition is a declarative sentence that is either true or false. Propositional attitudes include at least two components. First, there is the attitude itself. Hopes, fears, dreads, wishes, thoughts, etc., are all different attitudes, different states of consciousness, and they are all different from each other based on their conscious feel. A hope is a different form of consciousness from an episode of fear. A hope that it will rain is different from a fear that it will rain. What's the difference? A hope has a very different conscious feel from a fear.

Second, they all have a content or a meaning embedded in a proposition—namely, the propositional content of my consciousness while I am having the attitude. My hope that *P* (for example, that I am having eggs for breakfast) differs from my hope that *Q* (say, that it won't rain today) because *P* and *Q* are different propositions or meanings in my consciousness, even though the attitude (hoping) is the same in each case. My hope that it will rain is different from my hope that taxes will be cut. The contents of these two hopes have quite different meanings. If there were no conscious selves, there would be no propositional attitudes.

A third type of mental entity is *acts of free will or purposings.* What is a purposing? If, unknown to me, my arm is tied down and I still try to raise it, then the purposing is the "trying to bring about" the event of raising my arm. Intentional actions are exercises of active power by conscious selves wherein and whereby they do various things. They are free acts of will performed by conscious selves.

To summarize, dualists argue that sensations, propositional attitudes, and purposings are all examples of mental entities.

In addition to these differences between physicalists and dualists, there is also an intramural debate between *mere *property dualists* and **substance dualists.*

Mere property dualists believe there are some physical substances that have only physical properties: For example, a billiard ball being hard and round. They also maintain that there are no mental substances. On the other hand, they contend there is one material substance that has both physical *and* mental properties—the brain. When I experience a pain, there is a certain physical property possessed by the brain (a C-fiber stimulation with chemical and electrical properties) and there is a certain mental property possessed by the brain (the pain itself with its felt quality). The C-fiber event may cause the pain event, but they are two events, not one. The brain is the possessor of all mental properties and events. I am not a mental self that *has* my thoughts and experiences. Rather, I am a brain and a series or bundle of successive experiences themselves. Moreover, property dualists claim that, just as wetness is a real property that *supervenes or emerges upon a group of water molecules, so mental properties supervene/emerge upon brain states.

In contrast with property dualism, substance dualism holds that the brain is a physical thing that has physical properties, and the mind or soul is a mental substance that has mental properties. When I am in pain, the brain has certain physical properties (electrical, chemical) and contains certain physical states (e.g., C-fiber firing events), and the soul or self has certain mental properties (the conscious awareness of pain) and contains certain mental events (a pain state, an episode of thinking). The soul is the possessor of its experiences. It stands behind, over, and above them and remains the same throughout my life. The soul and the brain can interact with each other, but they are different entities with different properties. While in the body, the soul's functioning may depend on the proper working of the brain or other organs (e.g., the eyes). Since the soul is not to be identified with any part of the brain or with any particular mental experience, the soul may be able to survive the destruction of the body. Substance dualists accept the existence of both mental properties and substances. So substance dualists are also property dualists (they believe consciousness is a mental property), but substance dualists are not *mere* property dualists (those who deny a spiritual soul or self).

THE NATURE OF IDENTITY

It is time to turn to a topic that will explain our strategy for defending property and substance dualism: *the nature of identity*. The eighteenth-century philosopher/theologian Joseph Butler once remarked, allegedly, that everything is itself and not something else. This simple truth has profound implications. Suppose you want to know whether J. P. Moreland is Eileen Spiek's youngest son. If J. P. Moreland is identical to Eileen Spiek's youngest son, then in reality, we are talking about one single thing: J. P. Moreland, who *is* Eileen Spiek's youngest son. And everything true of J. P. Moreland will be true of Eileen Spiek's youngest son, and vice versa. However, if even one small thing is true of J. P. Moreland and *not* true of Eileen Spiek's youngest son, then these are two entirely different people. Furthermore, J. P. Moreland is identical to himself and not different from himself. So if J. P. Moreland is *not* identical to Eileen Spiek's youngest son, then in reality we must be talking about two things, not one.

This illustration suggests a truth about the nature of identity known as *Leibniz's Law of the Indiscernibility of Identicals* (from the German philosopher G. W. Leibniz who formulated it): For any entities x and y, if x and y are identical (they are really the same thing, there is only one thing you are talking about, not two), then any truth that applies to x will apply to y as well. This suggests a test for identity: If you could find one thing true of x not true of y, or vice versa, then x cannot be identical to (be the same thing as) y. Further, if you could find one thing that could *possibly* be true of x and not of y (or vice versa), even if it isn't actually true, then x cannot be identical to y.

For example, if J. P. Moreland is five feet and eight inches tall, but Eileen Spiek's youngest son is six feet tall, then they are not the same thing. Further, if J. P. Moreland is five feet eight and Eileen Spiek's youngest son is five feet eight, but it would be possible for J. P. to be five feet nine while Eileen's youngest son were five feet ten, then they are not the same thing either.

What does this have to do with the mind/body problem? Simply this: Physicalists are committed to the claim that alleged mental entities—substances, properties, events/states—are really identical to physical entities, such as brain states, properties of the brain, overt bodily behavior, and dispositions to behave (for example, pain is just the tendency to shout "Ouch!" when stuck by a pin, instead of pain being a certain mental feel of hurtfulness). If physicalism is true, then everything true of the brain (and its properties, states, and dispositions) is true of the mind (and its properties, states, and dispositions) and vice versa.[2] If we can find one thing true, or even possibly true, of the mind and not of the brain, or vice versa, then dualism is established. Then the mind or its properties and states is not the brain or its properties and states.

In some of the chapters to follow, I will present a number of arguments that imply that something is true of the mind or its states and not the brain or its states, or vice versa, and thus the former cannot be identical to the latter. But if they are not identical, physicalism is false and, taking dualism to be the only other option, dualism would be true.

WHY THE FINDINGS OF NEUROSCIENCE ARE LARGELY IRRELEVANT TO THE DEBATE

Keep in mind that the relation of identity is different from any other relation—for example, the relation of causation or constant connection. It may be that brain events cause mental events or vice versa: Having certain electrical activity in the brain may cause me to experience a pain; exercising an intention to raise my arm may cause bodily events. It may be that for every mental activity, a neurophysiologist can find a physical activity in the brain with which it is correlated. But just because A causes B (or vice versa), or just because A and B are constantly correlated with each other, that does not mean that A is *identical to* B. Sunlight may cause me to sneeze, but it's clear that the sunlight is not the same thing as my sneezing. Something is trilateral (three sided) if and only if it is triangular (three angled). But trilaterality (the property of having three sides) is not identical to triangularity (the property of having three angles), even though they are constantly conjoined.

Therefore, and this is critical, strict physicalism cannot be established by showing that mental states and brain states are interdependent on, causally related, or constantly conjoined with each other in an embodied person. *Physicalism needs identity to make its case, and if something is true, or possibly true of a mental substance, property, or event that is not true or possibly true of a physical substance, property, or event, then strict physicalism is false.*

For example, it is sometimes claimed that neuroscience has demonstrated that items such as memories are really just physical goings-on in certain regions of the brain. Now, what is the basis for such claims? The neuroscientist will attach certain probes, for example, an EEG, to various regions of the scalp and ask the subject to try not to think of much in order to establish a baseline reading of the electrical activity in various regions of the subject's brain. Then the scientist will present a series of numbers to the patient and, occasionally, interrupt the series and ask him to recall the number that was two numbers removed from the currently presented number. While the subject is engaging in this act of memory, the neuroscientist records increased electrical activity in certain regions of the brain and concludes that memories just are those activities. However, it should be clear that all that has been established is a correlation, not an identity, between the mental act of remembering and the activated network of brain firing. In general, neuroscience is wonderful for providing information about the neurological aspects of mental functioning and the self's actions, but it is of no help whatsoever in telling us what mental states and the self are. Correlation, dependence, and causal relations are not identity.

We should have known this all along, and it becomes evident when we observe that certain leading neuroscientists—Nobel Prize winner John Eccles, U.C.L.A. neuroscientist Jeffrey Schwartz, and Mario Beaureguard—are all dualists and they know the neuroscience.[3] Their dualism, and the central intellectual issues involved in the debate, are quite independent of neuroscientific data. As we shall see in later chapters, those issues are largely theological and philosophical, not scientific.

The irrelevance of neuroscience also becomes evident when we consider the recent bestseller *Proof of Heaven* by Eben Alexander.[4] Regardless of one's view of the credibility of near-death experiences (NDEs) in general, or of Alexander's in particular, one thing is clear: before whatever it was that happened to him, Alexander believed the standard neuroscientific view that specific regions of the brain generate and possess specific states of consciousness. But after his NDE, Alexander came to believe that it is the soul that possesses consciousness, not the brain, and the various mental states of the soul are in two-way causal interaction with specific regions of the brain. Here's the point: his change in view was a change in metaphysics that did not require him to reject or alter a single neuroscientific fact in which he believed. Dualism and physicalism are empirically equivalent views consistent with all and only the same scientific data. Thus, the authority of empirical data in science cannot be claimed on either side.

CHAPTER IN REVIEW

In this section I introduced a number of concepts that are crucial for understanding the mind/body question, such as substances, properties, and events. You will want to familiarize yourself with these, and other significant terms discussed in the chapter (see Key Vocabulary below). We also discussed a number of important differences between physicalism and dualism, and contrasted physical properties with mental properties. Additional points of importance include the following:

- **According to Leibniz's Law of the Indiscernibility of Identicals: For any entities x and y, if x and y are identical, then any truth that applies to x will apply to y as well.**
- We can use Leibniz's Law to show that something is true of the mind or its states and not the brain or its states, demonstrating that physicalism is false and dualism, provided it is the only other option, is true.

- **The key issues are theological and philosophical and not neuroscientific.**
- Neuroscience shows *correlation* between mind and

brain, not that mind and brain are identical.

- Near-death experiences (NDEs) offer strong evidence that the soul possesses consciousness, not the brain, and the various mental states of the soul are in two-way causal interaction with specific regions of the brain.
- Dualism and physicalism are empirically equivalent views consistent with all and only the same scientific data. Thus, the authority of empirical data in science cannot be claimed on either side.

KEY VOCABULARY

Event: A temporal state that occurs in the world (e.g., water freezing or a dog barking).

Knowledge: To represent reality in thought or experience the way it really is on the basis of adequate grounds.

Mind-body problem: The problem of understanding the relationship between the apparently immaterial mind and the physical body and brain.

Physicalism (or strict physicalism): The view that the only things that exist are physical substances, properties, and events. In relation to humans, the physical substance is the material body, especially the brain and central nervous system.

Property: An existent reality that is universal, immutable, and can (or perhaps must) be *in* or *had* by other things more basic, such as a substance. Thus, a cow (a substance) can have the property of being brown. The brownness (property) is had by the cow (the substance).

Property dualism: A human being is one material substance that has both physical *and* mental properties, with the mental properties arising from the brain.

Proposition: A declarative sentence that is either true or false. Examples of propositions include: "The earth orbits the sun," "Greg is six feet tall," and "I lived in Canada when I was seven."

Propositional attitude: An attitude (such as hoping, fearing, wishing, regretting) toward a certain proposition. For example, "I hope that the test will be cancelled," or "I fear that the economy is slowing down," or "I regret that I didn't have a second piece of cake."

Substance: a particular, individual, continuant and basic, fundamental existent thing that is a unity of parts, properties, and capacities, and has causal powers.

Substance dualism: A human person has both a brain that is a physical thing with physical properties and a mind or soul that is a mental substance and has mental properties.

Supervenience: A relationship of dependence between properties such that one level of the properties correlates to conditions at a different level. For example, when water molecules come together, the property of wetness supervenes upon them. In mind/body discussions, some philosophers (such as certain property dualists) hold that mental events supervene upon (or emerge from) brain events.

NOTES

1. Weak physicalism allows for supervenient "mental" properties as long as they are nomologically or metaphysically necessitated by their subvenient bases. There are two main ways to cash out this (non-reductive physicalist) view. First, physicalist functionalism according to which mental properties are functional properties with only physical realizers. This view cannot handle adequately the intrinsic nature of intentionality or phenomenal consciousness. Second, there is property dualism. The problem with this second alternative is that, once we grant genuine mental properties, we have strong intuitions that they are contingently related to their subvenient bases and this violates the necessitation requirement.
2. Different physicalists identify the person with different material objects, e.g., the brain, a sub-region of the brain, the entire living organism, an atomic simple. I will continue to make reference to the brain because that is the view most generally found in popular culture. For more on this see Eric Olson, *What Are We?* (Oxford: Oxford University Press, 2007).
3. See John C. Eccles and Karl Popper, *The Self and Its Brain* (London: Routledge, reprint edition 1984); Jeffrey Schwartz, *The Mind and the Brain* (New York: ReganBooks, 2002); Mario Beauregard and Denyse O'Leary, *The Spiritual Brain* (New York: HarperOne, 2008).
4. Eben Alexander, *Proof of Heaven* (New York: Simon & Schuster, 2012).

Chapter Two: THE BIBLE ON THE SOUL AND CONSCIOUSNESS

Most lay Christians would be surprised to learn that, among contemporary Christian intellectuals, there is a widespread loathing for substance dualism (hereafter, just "dualism"). We are often told that biblical revelation depicts the human person as a holistic unity whereas dualism is a Greek concept falsely read into the Bible by many in the history of the church.[1] Christians, we are told, are committed to physicalism and the resurrection of the body, not to dualism and the immortality of the soul. In short, dualism is outdated, unbiblical, and incorrect. For many Christian scholars, the very idea that some form of anthropological dualism is correct and required by Christian teaching is out of the question. Thus, Wolfhart Pannnenberg asserts, "The distinction between body and soul as two . . . different realms of reality can no longer be maintained. . . . [T]he separation between physical and spiritual is artificial."[2]

In my view, this aversion to dualism is sustained largely by sociological factors, for example, a rejection of Greek philosophy and an overestimation of its influence on biblical exegesis, or by various confusions—for example, that dualism is a rival theory to the resurrection of the body, that it is incompatible with a holistic emphasis that treats humans as body/soul functional unities, or that science has somehow shown that dualism must be rejected. However widespread Christian physicalism is, I believe that it cannot be sustained by a careful exegesis of the biblical text. As I hope to show in this chapter, Holy Scripture clearly teaches some form of anthropological dualism.

Before we look at biblical exegesis, it is important to note a few preliminaries. First, an important issue in biblical teaching is the Bible's view of the intermediate state between death and final resurrection. Currently, there are three views of the intermediate state. The first view is the traditional *temporary disembodiment* position that I defend below: A person is (or has) an immaterial soul/spirit deeply unified with a body but which can enter a temporary intermediate state of disembodiment at death, however unnatural and incomplete it may be, while awaiting a resurrection body in the final state. This view is clearly dualistic in nature.

The second view is a physicalist one and is called the *extinction/re-creation* position: Persons are identical to properly functioning bodies (or brains), and when the body dies the person ceases to exist since the person is in some sense the same as his or her body. At the future, final resurrection, persons are re-created after a period of non-existence. Among other things, this view has difficulty handling biblical texts that affirm a conscious intermediate state between death and final resurrection.

The third view may be called the *immediate resurrection* position: At death, in some way or another, each individual continues to exist in a physical way. Among other things, this view has difficulty making sense of why there will be a future, general resurrection. If one already has (or is) a post-death body, what is the need for a future, general resurrection?

Space considerations forbid me from critiquing directly the extinction/re-creation and immediate resurrection positions. However, my treatment of key biblical passages, though unfortunately brief, will show why I believe in the biblical superiority of the temporary disembodiment position in contrast to the other two.

The second preliminary is this: We should interpret a biblical passage in light of what its original audience would have understood it to mean. Now it is widely acknowledged that people all over the world throughout the ages have been substance dualists who depict life after death as involving the departure of the soul. This belief predates the biblical texts we will examine. As Raymond Martin and John Barresi note:

> As we have seen, Christianity played a decisive role in connecting that concern [with the self and its ability to endure], as it played itself out in the West, with philosophical theorizing. But the concern itself did not begin with Christianity. It did not even begin with the Greeks or Hebrews, or even the Egyptians. Rather, it began much earlier, perhaps with the Neanderthals, who twenty thousand years ago, in what is now France, Israel, and China, left remains of their dead in shallow graves, with carefully arranged, small, uniformly sized stones around them. No one knows exactly what these

> stones meant to the Neanderthals who put them there. But since the stones were purposefully and carefully arranged, it seems likely that they meant something. Most anthropologists think that the stones were arranged to express the Neanderthal's belief that those who were buried in this fashion continued to live beyond their bodily deaths.[3]

Dualism is pervasive and ancient. Indeed, as research has shown, people don't have to be taught to be dualists like they must if they are to be physicalists. Little children are naturally dualists. Summing up the recent research in developmental psychology, Henry Wellman states that "young children are dualists: knowledgeable of mental states and entities as ontologically different from physical objects and real [non-imaginary] events."[4]

Thus, unless we have significant textual reasons to the contrary, we should interpret relevant biblical passages as affirming dualism. The burden of proof is on the Christian physicalist to show that the commonsense view is mistaken.

Here's the final preliminary remark: Just as the Bible does not explicitly teach a theory of truth but instead clearly presupposes the commonsense one (i.e., the correspondence theory of truth according to which a proposition is true just in the case it corresponds to reality in the manner specified by the proposition), the Bible typically assumes substance dualism. As we will see in our treatment of two key New Testament texts, there are a few places where the explicit intent of the text is to affirm that the soul continues to live without the body in a disembodied intermediate state.

More frequently, the Bible implicitly affirms the reality of the soul without attempting to teach its existence explicitly. For example, in Matthew 10:28, Jesus warns us not to fear those who can only kill the body; rather, we should fear Him who can destroy both body and soul. The primary purpose of this text is to serve as a warning and not to teach that there is a soul. But in issuing His warning, Jesus implicitly affirms the soul's reality. And on other occasions, the Bible just assumes the commonsense view. For example, when the disciples who saw Jesus walking on water (Matt. 14:26) thought they were seeing a spirit, the Bible is merely assuming that we all know that we are (or have) souls that can exist without the body.

With these preliminaries in mind, let us turn to an examination of the Scriptures.

OLD TESTAMENT TEACHING

The main emphasis in Old Testament theology is on the functional, holistic unity of a human being. But a thing—for example, a car—can function as a unit even though it contains a plurality of components (tires, spark plugs, drive shaft). Accordingly, the Old Testament depiction of the functional unity of human persons includes an *ontological duality of immaterial/material components such that the individual human being can live after biological death in a disembodied intermediate state while awaiting the future resurrection of the body. There are two main lines of argument for this claim: an analysis of Old Testament anthropological terms and of Old Testament teaching about life after death. Let us look at these lines of argument in order.

OLD TESTAMENT ANTHROPOLOGICAL TERMS

Biblical anthropological terms exhibit a wide range of meanings, and Old Testament terms are no exception to this rule. Perhaps the two most important Old Testament terms are *nephesh* (frequently translated "soul") and *ruach* (frequently translated "spirit").

NEPHESH

The term *nephesh* occurs 754 times in the Old Testament and is used primarily of human beings, though it is also used of animals (Gen. 1:20; 9:10; 24:30) and of God Himself (Judg. 10:16; Isa. 1:14).[5] When the term is used of God, it clearly does not mean physical breath or life. Instead, it refers to God as an immaterial, transcendent self, a seat of mind, will, emotion, etc. (cf. Job 23:13, Amos 6:8). According to *A Hebrew and English Lexicon of the Old Testament* by Brown, Driver, and Briggs, the term has three basic meanings: various figurative usages, the life principle, and the soul of man that "departs at death and returns with life at the resurrection."[6]

To expand on this, in some places *nephesh* refers to a body part, for example, the throat (Isa. 5:14) or the neck (Ps. 105:18), and it can even be used to refer to a dead human corpse (Num. 5:2; 6:11). It sometimes refers to a desire of some sort (e.g., for food or sex).

On other occasions, *nephesh* refers to either life itself (Lev. 17:11: "the life [*nephesh*] of the flesh is in the blood") or to a vital principle/substantial entity that makes something animated or alive (Ps. 30:3: "You have brought up my soul [*nephesh*] from Sheol"; cf. Ps. 86:13; Prov. 3:22: "So [wisdom and discretion] will be life [*hayyim*] to your soul [*nephesh*])."

Nephesh also refers to the seat of emotion, volition, moral attitudes, and desire/longing for God (Mic. 7:1; Prov. 21:10; Isa. 26:9; Deut. 6:5; 21:14).

Finally, there are passages in which *nephesh* refers to the continuing locus of personal identity that departs to a disembodied afterlife as the last breath ceases (Gen. 35:18; cf. Ps. 16:10; 30:3; 49:15; 86:13; 139:8; 1 Kings 17:21, 22; Lam. 1:1). Death and resurrection are regularly spoken of in terms of the departure and return of the soul. Indeed, the problem of necromancy throughout Israel's history (the practice of trying to communicate with the dead in Sheol; cf. Deut. 18:9–14; 1 Sam. 28:7–25) seems to presuppose the view that ancient Israel took people to continue to live conscious, disembodied lives after the death of their bodies.

It is sometimes said that in these and other contexts, *nephesh* is simply a term that stands proxy for the personal pronoun "I" or "me" and, as such, it simply refers to the person as a totality. One way of refuting this objection is to claim that, frequently, the term *nephesh* is used as a figure of speech known as a *synecdoche* in which a part is used to represent the whole (e.g., "All hands on deck!"). Thus, *nephesh* does not refer to a part of the person but to the person as a whole psychophysical unity.

In my view, there is no clear textual evidence for this claim and, in fact, it is question begging because it fails to take seriously the fact that it is in virtue of the *nephesh* and not the body per se that the individual human is a living, sentient being capable of various states of emotion, volition, and so on. Thus, even if certain passages use *nephesh* to refer simply to the whole person ("Bless the LORD, O my soul [*nephesh*]" in Ps. 103:1), it is the whole person as a unified center of conscious thought, action, and emotion—that is, an ensouled body—to which reference is being made. Further, in cases of synecdoche of part for whole, even though the whole may be the intended referent of the term, implicit in the employment of the figure of speech is an acknowledgment of the reality of the part. When someone says, "All hands on deck!" he may be referring to entire persons, but he does so by way of a part, hands, that exist and are literal components of the wholes of which they are parts. The same is true of the *nephesh* when it is used in a synecdoche of part for whole.

Another argument against the dualistic construal of *nephesh* has been raised by Hans Walter Wolff. Speaking of the Old Testament use of *nephesh* to refer to a principle of life that can depart or return, Wolff says,

> [W]e must not fail to observe that the nephesh is never given the meaning of an indestructible core of being, in contradistinction to the physical life, and even capable of living when cut off from that life. When there is a mention of the "departing" (Gen. 35: 18) of the nephesh from a man, or of its "return" (Lam. 1:11), the basic idea . . . is the concrete notion of the ceasing and restoration of the breathing.[7]

Unfortunately, Wolff gives no adequate argumentation for this claim. Indeed, the clear reading of texts like Genesis 35:18 or 49:15 imply that it is the *nephesh* as a substantial principle of life and ground of consciousness and personal identity that leaves, continues to exist after biological death, and that can return. This seems clearly how *nephesh* is used when it is employed to refer to God Himself. In my view, Wolff's inadequate understanding of a functionally holistic form of substance dualism is turning his exegesis into eisegesis.

Moreover, when the Old Testament speaks of blood atonement to redeem the soul (Lev. 17:11: "I have given [the blood] to you . . . to make atonement for your souls [*nephesh*]"), the soul cannot merely refer to physical breath or life alone. Man's soul transcends mere physical or biological life and thus it has a form of significant intrinsic value that goes beyond mere physical breath or a set of physical processes. Similarly, when the Old Testament contains injunctions for people to humble their souls (Lev. 16:29; 23:27), they were not being commanded to prostrate their physical bodies or their biological life. They were to experience grief and sorrow in their transcendent selves.

Finally, the term *nephesh* is always translated *psuchē* and never *bios* in the Septuagint, a Greek translation of the Old Testament during the intertestamental period. The term *bios* is the Greek word for mere biological or physical life and the regular avoidance of this term by the translators of the Septuagint is best explained by their recognition that *nephesh* refers to a transcendent, irreducible aspect of living things that goes beyond mere breath or physical life.

RUACH

The other key Old Testament term is *ruach*, frequently translated "spirit." The term occurs 361 times and the breakdown of some of the specific translations in the King James Version is as follows: the Spirit of God (105 times), angels (23 times), the spirit in man (59 times), the wind (43 times), an attitude or emotional state (51 times), mind (6 times), breath (14 times).[8] *A Hebrew and English Lexicon of the Old Testament* by Brown, Driver, and Briggs lists nine meanings for the term: (1) God's Spirit, (2) angels, (3) the principle of life in humans and animals, (4) disembodied spirits, (5) breath, (6) wind, (7) disposition or attitude, (8) the seat of emotions, and (9) the seat of mind and will in humans. Definitions 1, 2, and 4 clearly seem to have straightforward dualist implications. Definitions 3 and 7–9 do as well when we realize that, if dualist arguments are successful, the principle/seat of life and consciousness is a transcendent self or immaterial ego of some sort. *Ruach* clearly overlaps with *nephesh*. However, two differences seem to characterize the terms. First, *ruach* is overwhelmingly the term of choice for God (though it is also used of animals; cf. Eccl. 3:19; Gen. 7:22) and, second, *ruach* emphasizes the notion of power. Indeed, if there is a central thread to *ruach*, it appears to be "a unified center of unconscious (moving air) or conscious (God, angels, humans, animals) power."

Ruach often refers to the wind insofar as it is an invisible, active power standing at God's disposal (Gen. 8:1; Isa. 7:2). In this sense, the *ruach* of God hovers over the waters with the power to create (Gen. 1:2). The term also signifies breath (Job 19:17) or, more frequently, a vital power that infuses something, animates it, and gives it life and consciousness. In this sense, the *ruach* in man is given or formed by Yahweh (Zech. 12:1); it is that which proceeds from and returns to

Him and it is that which gives man life (Job 34:14). In Ezekiel 37, God takes dry bones, reconstitutes human bodies of flesh, and then adds a *ruach* to these bodies to make them living persons. Ezekiel 37 is parallel to Gen. 2:7 in which God breathes *neshamah*—a virtual synonym of *ruach* that means "the breath of life"—into an already formed body. In both texts, the entity God adds is (1) that which animates and makes alive, (2) something that is added by God and is non-emergent. The *ruach* is something that can depart upon death (Ps. 146:4; Eccl. 12:7; Job 4:15). There is no *ruach* in idols of wood or stone and thus they cannot move and do not possess consciousness (Hab. 2:19; Jer. 10:14).

Ruach also refers to an independent, invisible, conscious being as when God employs a spirit to accomplish some purpose (2 Kings 19:7; 22:21–23). In this sense, Yahweh is called the God of the spirits of all flesh (Num. 27:16; cf. 16:22). Here, "spirit" means an individual, conscious being distinct from the body. Moreover, *ruach* also refers to the seat of various states of consciousness, including volition (Deut. 2:30; Jer. 51:11; Ps. 51:10–12), cognition (Isa. 29:24), emotion (Judg. 8:3; 1 Kings 21:4), and moral/spiritual disposition (Eccl. 7:8; Prov. 18:14).

In light of our brief study of *nephesh* and *ruach,* it should be clear that belief in some form of Old Testament anthropological dualism is *prima facie* justified. Indeed, the burden of proof is on the physicalist, a burden made even more difficult when we turn to a direct examination of Old Testament descriptions of the intermediate state in Sheol.

THE OLD TESTAMENT ON LIFE AFTER DEATH

The Old Testament clearly depicts individual survival after physical death, however ethereal that depiction may be, in a form that seems to be disincarnate, that is, without flesh and bones. The dead in Sheol are called *rephaim*, sometimes translated "shades." As with most Old Testament terms, Sheol has a variety of meanings, including simply the grave itself. But there is no question that a major nuance of Sheol is a shadowy realm of all the dead (with the exception of Enoch and Elijah).

For a number of reasons, Old Testament teaching about life after death is best understood in terms of a diminished though conscious form of disembodied personal survival in an intermediate state. For one thing, life in Sheol is often depicted as lethargic, inactive, and in a way that resembles an unconscious coma (Job 3:13; Eccl. 9:10; Isa. 38:18; Ps. 88:10–12; 115:17–18). However, the dead in Sheol are also described as being with family, awake, and active on occasion (Isa. 14:9–10). Second, the practice of necromancy (communicating with the dead) is clearly taught as a real possibility and, on some occasions, an actuality (cf. Isa. 8:19; Lev. 19:31; 20:6; Deut. 18:11; 1 Sam. 28). Third, we have already seen that the *nephesh*—a conscious person without flesh and bones—departs to God upon death (cf. Ps. 49:15). Finally, the Old Testament clearly teaches the hope of resurrection beyond the grave (Job 19:25–27; Ps. 73:26; Dan. 12:2; Isa. 26:14, 19). It is possible to interpret these resurrection texts in a way that denies a conscious, intermediate state and we will look at this possibility shortly when we turn to the New Testament. However, it seems clear that the most natural way to interpret them is in terms of the soul/spirit as the locus of personal identity that survives death in a less than fully desirable disembodied state, and to which a resurrection body will someday be added.

The philosopher John Hick has objected to the "disembodied soul" view of the afterlife on the grounds that throughout primitive cultures, as well as in the Old Testament, the entity that survives in the afterlife is not an immaterial soul, but rather an ethereal surviving being; a shadowy, insubstantial, counterpart to the body; a quasi-bodily being.[9] In response, it must be admitted that the dead in Sheol are, in fact, depicted in language that might suggest a physical but shadowy figure, and that the "ethereal body" view cannot be ruled out absolutely.

But for at least four reasons, I think that the ethereal body view of the intermediate state is much weaker than the "disembodied soul" view: (1) The *nephesh* or *ruach* is viewed in Old Testament teaching as something that can depart at death, continue to exist, and return, and the *nephesh* or *ruach* seems clearly to be an immaterial, unifying locus of personal identity and ground of various mental and living functions. (2) Throughout Scripture, sensory imagery is used in a non-literal way to describe immaterial, invisible realities, including heaven and hell, angels and demons, and God Himself. In these cases, the visual imagery is not taken literally, especially in descriptions of spirits and God. Moreover, it is possible to understand the various visitations of angels, Moses and Elijah in the transfiguration, and the Angel of Yahweh (God Himself) as either temporary embodiment or the power to manifest sense-perceptible qualities without being physical.

(3) Old Testament teaching implies that the soul or spirit is added to flesh and bones to form a living human person (Gen. 2:7; Ezek. 37) and that the resurrection of the dead involves the re-embodiment of the same soul or spirit (Isa. 26:14, 19). This is more consistent with the disembodied soul view than with the ethereal body position. (4) John Cooper has shown that intertestamental Judaism clearly used *nephesh* and *ruach* to refer to deceased, immaterial persons in a disembodied intermediate state and that the best way to explain this usage is to see it as expanding on and clarifying ideas already contained in Old Testament teaching.[10]

In sum, the Old Testament teaches that the soul/spirit is an immaterial entity that grounds and unifies conscious, living functions; that constitutes personal identity; that can survive physical death in a diminished form in the intermediate state; and, eventually, be reunited with a resurrection body.

NEW TESTAMENT TEACHING

When we turn to the New Testament, this dualistic view of human persons becomes even more compelling. However, before we proceed, four preliminary remarks should be made.

First, I acknowledge that the New Testament does not attempt to develop a philosophical anthropology as its primary focus. It does not follow from this, however, that New Testament data do not provide sufficient evidence to rule out certain anthropological models such as physicalism, and to justify others, such as some form of substance dualism.[11] Second, I also acknowledge that certain New Testament texts use *psuchē* (soul) or *pneuma* (spirit) as a synecdoche of part for whole (cf. Luke 12:19: "And I will say to my soul, 'Soul, you have many goods laid up for many years to come; take your ease, eat, drink and be merry.'" Here, "soul" is used for the entire body/soul unity). Still, such figures of speech almost always imply the reality of the part (as "All hands on deck!" implies the reality of hands). Further, as we shall see shortly, there are clear texts where these terms are most naturally taken to refer to an immaterial self. Third, New Testament anthropological terms possess wide ranges of meaning and the precise usage of such terms should be determined on a text-by-text basis.

Finally, N. T. Wright has established that, while there were different views of the intermediate state in intertestamental Judaism, the Pharisees clearly held that at death the soul leaves the body, enters a disembodied state, and awaits the general resurrection of the body. This was the Pharisees' view in intertestamental Judaism, and Jesus (Matt. 22:23–33; cf. Matt. 10:28) and Paul (Acts 23:6–9; cf. 2 Cor. 12:1–4) side with the Pharisees on this issue over against the Sadducees.[12] More generally, in intertestamental Judaism, the intermediate state was widely understood as follows (cf. 1 Enoch 22:3, 4, 9; 2 Esdras 7:75, 78–80):[13] (1) The dead were clearly referred to as "souls" or "spirits," and these terms were widely employed to refer to disembodied persons. (2) The dead were considered conscious and active in the intermediate state. (3) Resurrection was depicted as the reunion of soul and body in a transformed, revivified bodily existence, though there were differences of opinion about the precise nature of the resurrection body.

As mentioned above, the Pharisees were among the groups that accepted 1–3 and they shaped the thinking of the common people in Jesus' day. The Sadducees appear to be the major exception to the rule, but opinion is divided about the precise nature of their beliefs. Some interpret them as believing in anthropological physicalism and annihilationism while others take them to have held to Old Testament teaching about Sheol in which the dead are cut off from God and usually unconscious and inactive.

These insights about intertestamental Judaism clearly place a burden of proof on anthropological physicalists since New Testament teaching ought to be interpreted in terms of what the original audience would have understood unless there is clear evidence to the contrary.

TWO EXPLICIT TEXTS FOR THE SOUL

In Matthew 22:23–33 and Acts 23:6–9, we have two texts whose explicit intent is to teach that Jesus and Paul, respectively, agreed with the Pharisees over against the Sadducees in affirming that at death the soul departs into a disembodied existence and awaits the general resurrection of the body. Let us look at these in order. Here is Matthew 22:23–33:

> On that day some Sadducees (who say there is no resurrection) came to Jesus and questioned Him, asking, "Teacher, Moses said, 'If a man dies having no children, his brother as next of kin shall marry his wife, and raise up children for his brother.' Now there were seven brothers with us; and the first married and died, and having

> no children left his wife to his brother; so also the second, and the third, down to the seventh. Last of all, the woman died. In the resurrection, therefore, whose wife of the seven will she be? For they all had married her."
>
> But Jesus answered and said to them, "You are mistaken, not understanding the Scriptures nor the power of God. For in the resurrection they neither marry nor are given in marriage, but are like angels in heaven. But regarding the resurrection of the dead, have you not read what was spoken to you by God: 'I am the God of Abraham, and the God of Isaac, and the God of Jacob'? He is not the God of the dead but of the living." When the crowds heard this, they were astonished at His teaching.

In this text, Jesus argues that if the Sadducees were correct, then Abraham, Isaac, and Jacob lived, died, and became extinct, and the text should have said, "I *was* the God of Abraham . . ." But it says, "I *am* the God of Abraham . . ."; that is, He continues to be their God because they continue to exist!

Now Christian physicalists can reply that they can hold a view logically consistent with this text by affirming that the patriarchs continued to exist as physical beings in the intermediate state. But, as Christians, our goal is not to develop views that are merely consistent with the Bible. In this case, we hunger to know what Jesus Himself believed and taught, and given the historical context, He and His audience clearly would have understood His statements to affirm a *disembodied* intermediate state.

Here is Acts 23:6–9:

> But perceiving that one group were Sadducees and the other Pharisees, Paul began crying out in the Council, "Brethren, I am a Pharisee, a son of Pharisees; I am on trial for the hope and resurrection of the dead!" As he said this, there occurred a dissension between the Pharisees and Sadducees, and the assembly was divided. For the Sadducees say that there is no resurrection, nor an angel, nor a spirit, but the Pharisees acknowledge them all. And there occurred a great uproar; and some of the scribes of the Pharisaic party stood up and began to argue heatedly, saying, "We find nothing wrong with this man; suppose a spirit or an angel has spoken to him?"

For present purposes, two things are important about this text. First, Paul sides with the Pharisees over against the Sadducees, as did Jesus. In verse 8, Paul affirms the Pharisees' doctrine of the resurrection of the dead, which implied "life after life after death," that is, death, followed by a disembodied intermediate state, followed by the general resurrection of the body. The emphasis here is not on the future resurrection, but on the disembodied intermediate state. Why? Because Paul is trying to persuade the audience that he has recently heard from someone in the afterlife (Jesus on the road to Damascus), and this would have readily been perceived by the Pharisees as a real possibility. Note their response in verse 9—maybe a spirit or an angel has spoken to Paul.

Second, in verse 8, "angel" does not refer to an angelic being, but to a departed person who now exists as a spirit. We know this because, among other things, if "angel" refers to an angelic being, then Luke is wrong about the Sadducees because they did, in fact, believe in angelic beings. And the term "spirit" refers to a human spirit or soul. In fact, the Greek phrase translated "nor an angel nor a spirit" seems to imply that the two words are simply different terms for the same thing—a disembodied human soul. Commenting on this verse, Ben Witherington III notes, "[I]t is reasonable to conclude that the term 'angel' (cf. I Enoch 22:3, 7; 103:3–4) or 'spirit' was sometimes used to refer to a deceased person. . . . This would suggest that 'angel or spirit' is seen as two different ways to refer to the same thing . . . "[14]

There are other New Testament texts in which the soul is *implicitly* affirmed. Thus, while it is not the central purpose of these texts to teach that there is a soul, they nevertheless affirm the soul's reality in expressing their central purpose.

NEW TESTAMENT ANTHROPOLOGY: NON-PAULINE

There are key non-Pauline New Testament passages that appear to use the term *spirit* in a dualistic sense.

1 Peter 3:18–20

For Christ also died for sins once for all, the just for the unjust, so that He might bring us to God, having been put to death in the flesh, but made alive in the spirit; in which also He went and made proclamation to the spirits now in prison, who once were disobedient, when the patience of God kept waiting in the days of Noah, during the construction of the ark, in which a few, that is, eight persons, were brought safely through the water.

In this text we are told that when Jesus was crucified, being alive in spirit He went and made proclamation to the spirits in prison who had been disobedient during the days of Noah. This text has two points of relevance for the anthropological debate. First, who are the spirits to whom Jesus preached? There are three main interpretations. Some argue that this text refers to the pre-incarnate Christ preaching to the wicked during the days of Noah. This interpretation is not likely, however, because it breaks with the chronological order of the passage: Jesus died (vs. 18), He preached (vs. 19), He ascended to heaven (vs. 22). Verse 18 contains two aorist participles in Greek (having been put to death, being made alive in the spirit) that present actions that occur at the same time as the main verb (Christ died), so the events described occurred at the time of the crucifixion. Interpretations two and three imply that between His death and resurrection, Christ preached to either disembodied spirits in the intermediate state or to imprisoned angels, respectively. The former view entails anthropological dualism, though the text is too ambiguous to allow dogmatism toward either interpretation.

The second point of relevance centers on Christ Himself. Between His death and resurrection, He continued to exist as a God-man in the intermediate state independently of His earthly body. Whatever it was about Jesus that allowed Him to continue to be a human could not be His earthly body. The most reasonable solution is that Jesus continued to have a human soul/spirit, a solution consistent with "made alive in the spirit" (vs. 18).

Hebrews 12:23

[But you have come] to the general assembly and church of the firstborn who are enrolled in heaven, and to God, the Judge of all, and to the spirits of the righteous made perfect . . .

The text refers to deceased but existent human beings in the heavenly Jerusalem as "the spirits of the righteous made perfect." "Spirits" is used to refer to human beings in either the intermediate state or after the final resurrection. Either way, deceased human beings are described as incorporeal spirits, a description fitting the context in which the heavenly Jerusalem is contrasted with what can be touched and empirically sensed (vv. 18–19). When this language is used of angels it clearly entails the idea of an angelic person who is identical to a substantial spirit and the same implication for human persons is most naturally seen in this text. Moreover, the verbs of Hebrews 12:18–24 are in the present tense, so it is highly probable that the verse is referring to disembodied persons in the intermediate state who await a final resurrection (cf. Heb. 11:35).

DEATH AS "GIVING UP THE SPIRIT"

Several texts refer to death as "giving up the spirit" (*pneuma* in Greek; Matt. 27:50; John 19:30; Luke 23:46; cf. 24:37–39; "gave up the ghost" is used in Acts 5:5, 10; 12:23 [KJV]). Most likely, this phrase expresses the idea of the departure of the person into the intermediate state and not simply the cessation of breathing because (1) Jesus committed Himself, not His breath, to God (Luke 23:46); (2) this was a standard way of referring to the disembodied dead in intertestamental Judaism (e.g., 2 Esdras 7:78: "Now concerning death, the teaching is: When the decisive decree has gone out from the Most High that a person shall die, as the spirit leaves the body to return again to him who gave it, first of all it adores the glory of the Most High."[NRSV]); (3) Luke 24:37–39 clearly uses "spirit" much like *rephaim* is used in the Old Testament—as a disincarnate person without "flesh and bones" (vs. 39).

There are also key non-Pauline New Testament passages that appear to use the term *soul* (Greek *psuchē*) in a dualistic sense. In Revelation 6:9–11, dead saints are referred to as the "souls" of the martyrs who are in the intermediate state awaiting the final resurrection (cf. Rev. 20:5–6). Here the intermediate saints are depicted as conscious and alive and are metaphorically described with sense-perceptible imagery in a way we have already described in our discussion of Old Testament imagery of *Sheol.*

Further, Matthew 10:28 says, "Do not fear those who kill the body but are unable to kill the soul; but rather fear Him who is able to destroy both soul and body in hell." In this text, *psuchē* seems clearly to refer to something that can exist without the body, and thus "soul" and "body" cannot simply be two different terms that refer to the person as a psycho-somatic unity. The most natural way to take Jesus' view here is to see it as an expression of a Jewish form of anthropological dualism. Some argue that this use of "soul" is a figure of speech and that the intent of the passage is simply to serve as a warning text. But, again, even if this is a synecdoche of part for whole (soul for the whole person), it still implies the reality of the part. Moreover, there is no textual evidence that this passage involves a figure of speech. Indeed, this sort of distinction between body and soul was used in intertestamental texts in a literal way. For example, in Testament of Job 20:3 we read: "Then the Lord gave me over into his hands to be used as he wished with respect to the body; but he did not give him authority over my soul."

NON-PAULINE TEACHING ON THE INTERMEDIATE STATE

A number of non-Pauline passages are most reasonably taken to affirm a disembodied intermediate state between death and final resurrection. In Jesus' debate with the Sadducees (Matt. 22:23–33; Mark 12:18–27; Luke 20:27–40), our Lord specifies the time of the resurrection as a general future event "in the age to come" (Luke 20:35), an understanding of the resurrection embraced by the Pharisees of that time and, as the context shows (Luke 20:39), they approve of Jesus' teaching about the intermediate state and resurrection. In John 5:28–29 and 11:23–24, Jesus also affirms that the final resurrection is a future event. Further, Jesus asserts that the patriarchs, as representatives of all people, are currently alive in the intermediate state because "all live to Him" (Luke 20:38). The passage in Matthew 22:32 clarifies this remark and shows that it does not mean that the patriarchs were alive to God's memory. In the Matthew passage, Jesus grounds his argument about the intermediate state in the continuous present tense of the verb He takes to be implicit in the Old Testament text He cites: God *is*, that is, *continues to be*, their God and thus they continue to be.

In addition, there is the transfiguration passage (cf. Matt. 17:1–13) in which Elijah (who never died) and Moses (who had died) appear with Jesus. The most natural way to interpret this text is to understand that Moses and Elijah have continued to exist—Moses was not re-created for this one event—and have been made temporarily visible. Thus, the transfiguration passage seems to imply an intermediate state, though, taken alone, it does not rule out a bodily view of persons in that state.

In the parable of Lazarus in Luke 16:19–31, we have a description of the intermediate state in Hades (not the final resurrection of the wicked in Gehenna). It is hard to know how far to press this parable, specifically, how much to make of the bodily, visual imagery in the text. But it seems safe to conclude from it that Jesus is at least teaching the existence of conscious, living persons in the intermediate state prior to the final resurrection.

In Luke 23:42–43, Jesus promises the thief on the cross that "today you shall be with Me in Paradise." The term "today" should be taken in its natural sense, namely, that the man would be with Jesus that very day in the intermediate state after their deaths. In intertestamental Judaism, paradise was sometimes taken to be the dwelling place of the faithful dead prior to the final resurrection (it could also be used of the final resurrection state). Now in Jesus' case, this text, coupled with other New Testament teaching on Christology, implies that Jesus continued to exist as a fully human person after His death and prior to His bodily resurrection. That is, He was a disembodied human soul with a full human nature united with a divine nature during the period between His death and resurrection. This would seem to imply that the thief existed in a disembodied intermediate state just like Jesus, which is possible only if the thief was more than his body.

NEW TESTAMENT ANTHROPOLOGY: PAULINE

When we turn to Pauline teaching, several strands of evidence unite to justify the claim that Paul taught a dualistic anthropology.

Acts 23:6–9

As we saw above, in this passage, Paul affirms his solidarity with the Pharisees over against the Sadducees in affirming the reality of angels, spirits, and the final resurrection. When Paul refers to his acceptance of the "resurrection of the dead," he means to affirm the Pharisaic teaching of the afterlife, which included the notion of the person as a disembodied spirit awaiting the final resurrection.

1 Thessalonians 4:13–18

Here Paul affirms the idea that at the return of Jesus, the dead in Christ shall be resurrected prior to those alive at that future time. This seems to teach clearly that individual deceased believers await a future, general resurrection. Moreover, Paul's description of those in the intermediate state as "asleep" simply describes persons who, while conscious and active, are not active in an earthly, bodily way. First Thessalonians 5:10 refers to those who are asleep as living together with Christ, a description that does not allow for an extinction/re-creation view of the afterlife.

1 Corinthians 15

This passage reaffirms the general teaching of 1 Thessalonians 4:13–18: There will be a general resurrection at the end of the age (cf. vv. 51–52) following a period of "sleep" (vv. 18, 20, 51)—a period of conscious, active, though diminished survival in a disembodied intermediate state. Moreover, verse 35 seems to make a clear distinction between persons and their bodies when Paul addresses the question of what sort of body the dead will have at the resurrection.

2 Corinthians 5:1–10 and Philippians 1:21–24

The traditional way of understanding 2 Corinthians 5:1–10 is as follows: Paul desires to live until the parousia (Greek, the second coming) because this would mean that he would have his earthly body immediately replaced with his resurrection body and thus he would not have to go through an unnatural condition of disembodiment in the intermediate state. Paul refers to the earthly body as the "earthly tent" (vs. 1), and he describes the resurrection body as a "building from God," a phrase that cannot refer to a heavenly dwelling place since it is something that can be put on (cf. vv. 2–3). Further, Paul refers to the disembodied intermediate state as a state of nakedness or of being unclothed (vv. 3–4), and he explicitly says that to be absent from the body is to be present with the Lord (vs. 8), thereby affirming the real possibility of disembodiment. If this interpretation is correct, then it has clear dualistic implications.[15]

Philippians 1:21–24 provides a parallel teaching to 2 Corinthians 5: Paul contrasts living in the body with temporary disembodiment with Christ in the intermediate state.

2 Corinthians 12:1–4

In 2 Corinthians 12:1–4 Paul describes a visionary experience he had lived through fourteen years earlier. In verse 3 he says that he does not know whether he was still in his body during the experience or whether it was a state of temporary disembodiment. Now it doesn't really matter for my argument which was correct. The simple fact that Paul allows for the possibility of his own temporary disembodiment is sufficient to show that he took himself to be non-identical to his body. It is because Paul understands himself as a soul/spirit united to a body that this was a real possibility for him.

Romans 8:18–23 and Philippians 3:20–21

In these two passages, Paul seems to affirm a future, general resurrection associated with the restoration of all things, a view that stands in stark contrast to an immediate personal resurrection position.

WHY DO BIBLICAL SCHOLARS REJECT SUBSTANCE DUALISM?

If my arguments in this chapter are correct, one might expect to find that all biblical scholars and theologians are substance dualists. But that would be far from the truth. In point of fact, many reject substance dualism. Why is this so? In my view, it is due to confusions about dualism on the part of biblical and theological scholars. As a paradigm case of such confusion, consider the writings of N. T. Wright. He has written that human persons are (or have) souls that are spiritual realities that ground personal identity in a disembodied intermediate state between death and final resurrection.[16] According to Wright, this was clearly the Pharisees' view in intertestamental Judaism, and, he notes, Jesus (Matt. 22:23–33) and Paul (Acts 23:6–9) side with the Pharisees on this issue over against the Sadducees. However, in a paper delivered in March 2011 at the Eastern Regional Meeting of the Society of Christian Philosophers, Wright explicitly disavowed dualism.[17] Yet, in the same paper, he affirms a dualist reading of 2 Corinthians 5:1–10, Acts 23:6–9, and 2 Corinthians 12:2–4 in keeping with his thesis that the Jews of Jesus' day, and the New Testament, affirm death followed by a *disembodied* intermediate state followed by the general resurrection.

Wright's confusion becomes evident when we distinguish generic dualism (the soul/mind/self is an immaterial substance that is different from the physical body) from radical Platonic dualism (the body is of little value and may, in fact, be evil; the soul is capable of immortal existence on its own steam without needing to be sustained by God; and disembodied existence is the ideal state in heaven with no need for a resurrected body). Wright is not careful to distinguish these, but it is the latter, not the former, that he rejects. I suggest a similar confusion characterizes much of the rejection of dualism on the part of biblical and theological scholars. For example, Christian physicalist Nancey Murphy seems guilty of this confusion. She states that in theological and biblical studies, there has been "a gradual displacement of a dualistic account of the person, with its correlative emphasis on the afterlife conceived in terms of the immortality of the soul."[18] Even if this is true, it follows only that radical Platonic dualism has been replaced, not that generic dualism is—or should be—replaced.

CHAPTER IN REVIEW

The following is a summary of the previous discussion of key Old and New Testament passages and concepts that illustrate the mind/body dualism taught in Scripture.

- **Preliminary Remarks**
- We should interpret a biblical passage in light of what its original audience would have understood it to mean, and substance dualism is an evident candidate.
- Just as the Bible does not explicitly teach a theory of truth but, instead, clearly presupposes the commonsense one, the Bible typically assumes substance dualism.
- The burden of proof is on the Christian physicalist to show that the commonsense view is mistaken.

- **Old Testament Teaching**
- The Old Testament depiction of the functional unity of human persons includes an ontological duality of immaterial/material components such that the individual human being can live after biological death in a disembodied intermediate state while awaiting the future resurrection of the body.
- In light of our brief study of *nephesh* and *ruach*, it should be clear that belief in some form of Old Testa-

ment anthropological dualism is *prima facie* justified.

- ☆ The *nephesh* or *ruach* is viewed in the Old Testament as something that can depart at death, continue to exist, and return; and the *nephesh* or *ruach* seems clearly to be an immaterial, unifying locus of personal identity and ground of various mental and living functions.
- ☆ Intertestamental Judaism clearly used *nephesh* and *ruach* to refer to deceased, immaterial persons in a disembodied intermediate state.
- The Old Testament clearly depicts individual survival after physical death, however ethereal that depiction may be, in a form that seems to be without flesh and bones.
- ☆ Old Testament teaching implies that the soul or spirit is added to flesh and bones to form a living human person (Gen. 2:7; Ezek. 37) and that the resurrection of the dead involves the re-embodiment of the same soul or spirit (Isa. 26:14, 19).
- ☆ The dead in Sheol are described as being with family, and awake and active on occasion (Isa. 14:9–10).
- ☆ The practice of necromancy is clearly taught as a real possibility and, on some occasions, an actuality (cf. Isa. 8:19; Lev. 19:31; 20:6; Deut. 18:11; 1 Sam. 28).
- The Old Testament teaches that the soul/spirit is an immaterial entity that grounds and unifies conscious, living functions; that constitutes personal identity; that can survive physical death in a diminished form in the intermediate state and, eventually, be reunited with a resurrection body.

- **New Testament Teaching**
- Although the New Testament does not develop a philosophical anthropology, the evidence for a dualistic view of human persons is quite compelling.
- Matthew 22:23–33 and Acts 23:6–10: Jesus and Paul agreed with the Pharisees that at death, the soul departs into a disembodied existence while awaiting the general resurrection of the body.
- New Testament Anthropology
  - ☆ 1 Peter 3:18–20: Jesus preached to either disembodied spirits in the intermediate state or to imprisoned angels, respectively. Moreover, between His death and resurrection, Jesus continued to exist as a God-man in the intermediate state independently of His earthly body.
  - ☆ 1 Thessalonians 4:13–18: Individual deceased believers await a future, general resurrection.
  - ☆ 2 Corinthians 5:1–10 and Philippians 1:21–24: The earthly body is like an "earthly tent," and the resurrection body like a "building from God" that can be put on. Further, the disembodied intermediate state is compared to a state of nakedness/being unclothed, and to be absent from the body is to be present with the Lord, thereby affirming the real possibility of disembodiment.
  - ☆ 2 Corinthians 12:1–4: Paul allows for the possibility of his own temporary disembodiment, which is sufficient to show that he took himself to be non-identical to his body.

KEY VOCABULARY

Extinction/Re-Creation View: When the body dies the person ceases to exist since the person is in some sense the same as his or her body. At the future, final resurrection, persons are re-created after a period of non-existence.

Immediate Resurrection View: At death, in some way or another, each individual continues to exist in a physical way.

Ontology (Ontological): A branch of metaphysics that deals with the nature of being and existence. Ontological questions include whether humans possess a soul, and whether abstract entities such as numbers truly exist.

Temporary Disembodiment View: A person is (or has) an immaterial soul/spirit deeply unified with a body that can enter a temporary intermediate state of disembodiment at death, however unnatural and incomplete it may be, while awaiting a resurrection body in the final state.

NOTES

1. For defenses of Christian physicalism, see Joel Green, *Body, Soul and Human Life* (Grand Rapids, MI: Baker Academic, 2008); Nancey Murphy, *Bodies and Souls, or Spirited Bodies?* (Cambridge: Cambridge University Press, 2006); Warren S. Brown, Nancey Murphy, and H. Newton Malony, eds., *Whatever Happened to the Soul?* (Minneapolis: Fortress, 1998).
2. Wolfhart Pannenberg, *What Is Man?* (Philadelphia: Fortress, 1970), 47–48.
3. Raymond Martin and John Barresi, *The Rise and Fall of Soul and Self: An Intellectual History of Personal Identity* (New York: Columbia University Press, 2006), 290.
4. Henry Wellman, *The Child's Theory of Mind* (Cambridge, MA: MIT, 1990), 50. I owe this reference to Stewart Goetz and Mark Baker.
5. See Robert A. Morey, *Death and the Afterlife* (Minneapolis: Bethany, 1984), 45–51.

6. Francis Brown, Samuel R. Driver, and Charles A. Briggs, *A Hebrew and English Lexicon of the Old Testament* (Oxford: Clarendon, 1953, 1977), 220.
7. Hans Walter Wolff, *Anthropology of the Old Testament* (Philadelphia: Fortress, 1974), 20.
8. See Morey, *Death and the Afterlife*, 51–53; Wolff, *Anthropology of the Old Testament*, 32–39.
9. John Hick, *Death and Eternal Life* (San Francisco: Harper & Row, 1976), 55–60.
10. John Cooper, *Body, Soul and Life Everlasting: Biblical Anthropology and the Monism-Dualism Debate* (Grand Rapids, MI: Eerdmans, 2000), 75–76, 81–103.
11. Murphy, *Bodies and Souls, or Spirited Bodies?*, 1–37.
12. N. T. Wright, *The Resurrection of the Son of God* (Minneapolis: Fortress, 2003), 131–34, 190–206, 366–67, 424–26.
13. Cf. Cooper, *Body, Soul and Life Everlasting*, 81–103. 1 Enoch 22:3, 4, 9 (300–100 BC): "Then Raphael answered, one of the holy angels who was with me, and said unto me: 'These hollow places have been created for this very purpose, that the spirits of the souls of the dead should assemble therein, yea that all the souls of the children of men should assemble here. And these places have been made to receive them till the day of their judgment and till their appointed period [till the period appointed], till the great judgment (comes) upon them. . . .' And he answered me and said unto me: 'These three have been made that the spirits of the dead might be separated. And such a division has been made (for) the spirits of the righteous, in which there is the bright spring of . . .' (http://www.ccel.org/c/charles/otpseudepig/enoch/ENOCH_1.HTM) 2 Esdras 7:75, 78–80 (AD 95–200): "I answered and said, 'If I have found favor in thy sight, O Lord, show this also to thy servant: whether after death, as soon as every one of us yields up his soul, we shall be kept in rest until those times come when thou wilt renew the creation, or whether we shall be tormented at once?' . . . Now, concerning death, the teaching is: When the decisive decree has gone forth from the Most High that a man shall die, as the spirit leaves the body to return again to him who gave it, first of all it adores the glory of the Most High. And if it is one of those who have shown scorn and have not kept the way of the Most High, and who have despised his law, and who have hated those who fear God—such spirits shall not enter into habitations, but shall immediately wander about in torments, ever grieving and sad, in seven ways." (http://www.biblestudytools.com/rsva/2-esdras/passage.aspx?q=2-esdras+7:71-81)

14. Ben Witherington III, *The Acts of the Apostles* (Grand Rapids, MI: Eerdmans, 1998), 692. Cf. p. 387.
15. For an alternative interpretation favoring the immediate resurrection view, see Murray Harris, *Raised Immortal: Resurrection & Immortality in the New Testament* (Grand Rapids, MI: Eerdmans, 1985), 219–26. For a response, see John Cooper, *Body, Soul and Life Everlasting*, 155–63.
16. See N. T. Wright, *The Resurrection of the Son of God* (Philadelphia: Fortress, 2003), 131–34, 190–206, 366–67, 424–26.
17. N. T. Wright, "Mind, Spirit, Soul and Body: All for One and One for All: Reflections on Paul's Anthropology in His Complex Contexts," presented at the Society of Christian Philosophers Regional Meeting, Fordham University, March 18, 2011.
18. Murphy, *Bodies and Souls, or Spirited Bodies?*, 10.

Chapter Three: THE NATURE AND REALITY OF CONSCIOUSNESS

*__Consciousness is among__ the most mystifying features of the cosmos. Philosopher Colin McGinn claims that its arrival borders on sheer magic because there seems to be no naturalistic explanation for it: "How can mere matter originate consciousness? How did evolution convert the water of biological tissue into the wine of consciousness? Consciousness seems like a radical novelty in the universe, not prefigured by the after-effects of the Big Bang; so how did it contrive to spring into being from what preceded it?"[1]

I believe that the existence of God is the best explanation for finite examples of consciousness in creatures such as humans and various animals. Finite consciousness provides strong evidence that God exists. In my view, scientific naturalism is utterly incapable in principle of providing any explanation whatever for finite consciousness. The scientifically, indeed, the culturally authorized naturalist story of how all things came about revolves around the atomic theory of matter and evolutionary theory. As Phillip Johnson observes, "The materialist story is the foundation of all education in all the departments in all the secular universities, but they do not spell it out. It is,

> In the beginning were the particles and the impersonal laws of physics.
> And the particles somehow became complex living stuff;
> And the stuff imagined God;
> But then discovered evolution."[2]

According to the atomic theory of matter, all chemical change is the result of the rearrangement of tiny parts—for example, protons, neutrons, and electrons. According to evolutionary theory, random mutations are largely responsible for providing an organism with a change in characteristics; some of those changes provide the organism with a survival advantage over other members of its species; as a result, the organism's new traits eventually become ubiquitous throughout the species.

This story is physically deterministic in two ways. First, the physical state of the universe (and everything in it, including you) at a particular time and the impersonal laws of nature are sufficient to determine or fix the chances of the next successive state. This is *temporal* determination. Second, the features and behavior of ordinary sized objects like glaciers, rocks, human beings, and animals is fixed by the states of their atomic and subatomic parts. This is *bottom-up* or *parts-to-whole* determinism. On naturalism, if genuinely mental consciousness exists, it is a causally impotent *epiphenomenon. An epiphenomenon is something that is caused to exist by something else but that itself has no ability to cause anything. For example, fire causes smoke, but smoke, we may assume for the purposes of illustration, does not, in turn, cause anything. Smoke is an epiphenomenon. Among other things, this means that a feeling of thirst never causes someone to get a drink—thoughts and beliefs play no role in directing or bringing about our behavior, since they are causally impotent. Many philosophers rightly think that if a view implies *epiphenomenalism, the view must be rejected.

Naturalism is also a strictly physical story. And that is where a second, and most fundamental problem of consciousness enters the picture. If you begin with matter and simply rearrange it according to physical laws by means of strictly physical causes and processes, then you will end up with increasingly different arrangements of—you guessed it—matter. Start with matter and tweak it physically and all you will get is tweaked matter. There is no need or room for mind and consciousness to enter the picture. However, if you begin with God, then Mind is the fundamental reality (not matter) and its appearance in cosmic history is not the ontological problem it is for the scientific naturalist. But we are getting ahead of ourselves.

THE NATURE OF CONSCIOUSNESS

While the *origin* of consciousness is surrounded by controversy, the *nature* of consciousness is pretty commonsensical. Suppose you are in the recovery room immediately after surgery. You are still deeply under anesthesia. Suddenly and somewhat faintly, you begin to hear sounds. It is not long until you can distinguish two different voices. You begin to feel a dull throb in your ankle. The smell of rubbing alcohol wafts past your nose. You remember a childhood accident with the same smell. You feel an aversion toward it. You feel thirsty and desire a drink. As you open your eyes to see a white ceiling above, you begin to think about getting out of the hospital. What is going on? The answer is simple: You are regaining consciousness.

Note two things about this example. First, whereas any physical object (state, process, property, relation) can (and only can) be completely described from a third-person perspective, descriptions of states of consciousness require a first-person point of view (only an "I" can describe or report them; compare this to describing, say, a hospital bed, which we can only observe from the "outside."). Second, states of consciousness are best defined by citing or pointing to specific examples. In fact, a fairly good characterization of consciousness is this: Consciousness is what you are aware of when you engage in first-person introspection. Both of these observations are exactly what the dualist approach to consciousness would predict.

At least five kinds of conscious states exist (see also the discussion of mental entities in chapter 1). A *sensation* is a state of awareness or sentience, e.g., a conscious awareness of sound or pain. Some sensations are experiences of things outside me like a tree or the color red. Others are awarenesses of states within me like pains or itches. Emotions are a subclass of sensations and, as such, they are forms of awareness of things. I can be aware of something in an angry or sad way. A *thought* is a mental content that can be expressed in a sentence. Some thoughts logically imply other thoughts. For example, "All dogs are mammals" entails "This dog is a mammal." If the former is true, the latter must be true. Some thoughts don't entail, but merely provide evidence for other thoughts. For example, certain thoughts about evidence in a court case provide grounds for the thought that a person is guilty. Thoughts are the sorts of things that can be true or false, reasonable or unreasonable. A *belief* is a person's view, accepted to varying degrees of strength, of how things really are. A *desire* is a certain felt inclination to do, have, or experience certain things or to avoid them. An *act of will* is a choice, an exercise of power, an endeavoring to act, usually for the sake of some purpose.

A CASE FOR PROPERTY DUALISM

Are properties such as a pain or a thought, and the states/events composed of them (a pain or thinking event) genuinely mental, or are they physical? In this chapter, we will look at a case for property dualism, the view that conscious properties/events are mental and not physical. In order to follow the case, remember the law of identity that we looked at in chapter 1: If x is identical to (is the same thing as) y, then whatever is true of x is true of y, and vice versa. If we can find one thing true or possibly true of x not true or possibly true of y, they are not the same thing, even if one of the two depends on the other for its functioning.

Property dualists argue that mental states are in no sense physical since they possess *five* features not owned by physical states:

(1) there is a raw qualitative feel or a "what it is like" to having a mental state such as a pain (e.g., we can

easily tell a pain from a feeling of joy, since the two experiences are qualitatively different);

(2) many mental states have *intentionality—*of*-ness or *about*-ness directed towards an object (e.g., I can have a thought *about* a cat or *of* a lake);

(3) mental states are inner, private, and immediate to the subject having them;

(4) mental states require a subjective ontology—that is, mental states are necessarily owned by the first-person subjects who have them (only *I* can possess my thoughts; no one else can);

(5) mental states fail to have crucial features (e.g., spatial extension, location, being composed of parts) that characterize physical states and, in general, cannot be described using physical language (my thoughts have no physical dimensions, no physical location, and aren't made of simpler building blocks).

Space considerations forbid me from undertaking a defense of the mental nature of all five of these features. Rather, with these points in mind, we shall examine three important arguments for property/event dualism.

THE INTROSPECTIVE DIFFERENCE BETWEEN CONSCIOUSNESS AND THE PHYSICAL

First, once one gets an accurate description of consciousness (see above), it becomes clear that mental properties/events are not identical to physical properties/events. Mental states are characterized by their intrinsic, subjective, inner, private, qualitative feel, made present to a subject by first-person introspection. For example, a pain is a certain felt hurtfulness. The intrinsic nature of mental states cannot be described by physical language, even if, through study of the brain, one can discover the causal/functional relations between mental and brain states.

In general, mental states have some or all of the following features, none of which is a physical feature of anything: Mental states like pains have an intrinsic, raw, conscious feel. There is a "what-it-is-like" to a pain. But there isn't a similar "what-it-is-like" to physical states like boiling at a certain temperature or existing as a liquid. Most, if not all, mental states have intentionality—they are *of* or *about* things. But no physical state is *of* or *about* something. A thunderstorm, for example, isn't *about* or *of* anything. Mental states are inner, private, and known by first-person, direct introspection. But any way one has of knowing about a physical entity is available to everyone else, including ways of knowing about one's brain. You and I can both look at a CT scan of my brain and discuss if everything looks normal. But only I can know what I'm thinking. Mental states—for example, a pain event—contain simple qualities (a feeling of hurtfulness) that aren't composed of parts. A pain in the foot is just a simple felt hurtfulness; the pain doesn't have parts. Indeed, it is hard even to understand what it would mean to say that the pain *did* have parts. By contrast, the brain states associated with these simple mental states, for example, a neuronal firing event called a C-fiber firing event associated with a pain event, is, in fact, composed of millions of parts: atoms, molecules, cells, and so forth. Being a C-fiber firing event or being a neural network synchronous firing event is a complex event, but a sensation or thought is a simple event.

Mental states are constituted by *self-presenting properties, but physical states are not. What does this mean? One can be aware of the external, physical world only by means of one's mental states (one is aware of a lamp by means of having a sensation of the lamp), but one need not be aware of one's mental states by means of anything else (I am aware of my sensation of the lamp by simply having that sensation itself; I may, but need not, have a second-order awareness of my sensation if I choose to focus on that sensation and not on the lamp). Again, it is by way of a sensation of red that one is aware of an apple, but one is not aware of the sensation of red by way of another sensation. Mental states are necessarily owned, and, in fact, one's mental states could not have belonged to someone else. However, no physical state is necessarily owned, much less necessarily owned by a specific subject. I necessarily own my sensation of red, but cannot own the physical state of being red.

Some sensations are vague—for example, my sensation of a tree may be fuzzy or vague—but no physical state is vague. Some sensations are pleasurable or unpleasant, but nothing physical has these properties. A cut in the knee is, strictly speaking, not unpleasant. It is the pain event caused by the cut that is unpleasant. Mental states can have the property of familiarity (e.g., when a desk looks familiar to someone), but familiarity is not a feature of a physical state.

Since mental states have these features and physical states do not, then mental states are not identical to physical states. Some physicalists have responded by denying that consciousness has the features in question. For example, dualists have argued that thinking events are not spatially located even though the brain event associated with them is. Physicalists counter that thoughts are, after all, located in certain places of the brain. But there is no reason to accept this claim, since dualists can account for all the spatial factors in terms of the brain events causally related to thoughts. It is the brain events associated with thoughts that have spatial features (size, shape, location), features that can be quantified. But it is not the thinking events that have these spatial features.

Moreover, through introspection subjects seem to know quite a bit about the features of their thoughts, and spatial location is not one of them. Similar responses are offered by dualists for physicalist claims about the other features of consciousness.

THE KNOWLEDGE ARGUMENT

A second argument for property/event dualism is the Knowledge Argument, variously formulated by Thomas Nagel, Frank Jackson, and Saul Kripke.[3] A standard presentation of the thought experiment describes Mary, a brilliant scientist blind from birth, who knows all the physical facts relevant to acts of perception. When she suddenly gains the ability to see, she gains knowledge of new facts. Since she knew all the physical facts before recovery of sight, and since she gains knowledge of new facts, these facts must not be physical facts and, moreover, given Mary's situation, they must be mental facts.

To appreciate the argument, it is necessary to focus on

the nature of self-presenting properties and three kinds of knowledge. First, a self-presenting property, such as *being appeared to redly* (that is, experiencing an appearance of the color red), presents both its intentional object (say, a red apple) and itself (the redness) to the subject experiencing it. When a person is sensing an apple, they are aware of the apple and they are also having the experience of seeing the apple. Second, arguably, there are three forms of knowledge, irreducible to each other (though, of course, one form may be the epistemic ground for another): (1) **Knowledge by acquaintance*: One has such knowledge when one is directly aware of something. For example, when one sees an apple directly before him, he knows it by acquaintance. One does not need a concept of an apple or even knowledge of the word "apple" to have knowledge by acquaintance of an apple. (2) **Propositional knowledge*: This is knowledge that a proposition is true. For example, knowledge that "the object over there is an apple" requires having a concept of an apple and knowing that the object under consideration satisfies the concept. (3) **Know-how*: This is the ability to do certain things, say, to use apples to make pies.

Generally, knowledge by acquaintance provides grounds for propositional knowledge, which, in turn, provides what is necessary to have genuine know-how. It is because one sees the apple that one knows that it is an apple, and it is in virtue of one's knowledge of apples that one has the skill to do things to or with them.

By way of application, Mary comes to have the self-presenting mental property of being appeared to redly (she experiences a sensation of redness). In this way, Mary gains six new kinds of knowledge—knowledge by acquaintance, propositional knowledge, and skill both with regard to the color red and her sensation of red. Mary now knows by acquaintance what redness is. Upon further reflection and experience, she can now know things like "Necessarily, red is a color." She also gains skill about comparing or sorting objects on the basis of their color, of how to arrange color patterns that are most beautiful or natural to the eye, etc. Assuming a realist (and not a representative) construal of secondary qualities (secondary qualities are colors, sounds, tastes, smells, textures, and a realist view of them implies that objects in the world have these features; so when a tree falls in the forest with no one around, there is, indeed, a sound!), we may say that the three kinds of knowledge just listed are not themselves knowledge of facts about the brain, but are forms of knowledge that can be gained only by way of mental states that exemplify the relevant self-presenting property (only by having a sensation of red can Mary have these forms of knowledge about redness).

Further, Mary gains knowledge about her sensation of red. She is now aware of having a sensation of red for the first time and can be aware of a specific sensation of red being pleasurable, vague, etc. She also has propositional knowledge about her sensations. She could know that a sensation of red is more like a sensation of orange than it is like a sour taste. She can know that the way the apple appears to her now is vivid, pleasant, or like the way the orange appeared to her (namely, redly) yesterday in bad lighting. Finally, she has skill about her sensations. She can recall them to memory, re-image things in her mind, adjust her glasses until her sensations of color are vivid, etc.

Physicalists David Papineau and Paul Churchland have offered slightly different versions of the most prominent physicalist rejoinder to this argument:[4] When Mary gains the ability to see red, she gains no knowledge of any new facts. Rather, she gains new abilities, new behavioral dispositions, new know-how, new ways to access the facts she already knew before gaining the ability to see. Before being able to see, Mary knew all there was to know about the physical facts involved in what it is like to experience red. She could third-person imagine what it would be like for some other person to experience red. She could know what it is like to have an experience of red due to the fact that this is simply a physical state of the brain and Mary had mastered the relevant physical theory before gaining sight. But now she has a "*pre-linguistic representation* of redness," a *first-person* ability to *image* redness or *recreate* the experience of redness in her memory. She can *re-identify* her *experience* of red and *classify* it according to the type of experience it is by a new "*inner" power of introspection*. Prior to the experience of seeing, she could merely recognize when someone else was experiencing red "from the outside," that is, from observing the behaviors of others. Thus, the physicalist admits a duality of types of knowledge but not a duality of facts that are known.

For three reasons, this response is inadequate. First, it is simply not true that Mary gains a new way of knowing what she already knew instead of gaining knowledge of a new set of facts. We have already listed some of Mary's new factual knowledge, and it seems obvious that Mary failed to have this factual knowledge prior to gaining the ability to see.

Second, to be at all plausible, this physicalist rejoinder seems to presuppose a course-grained theory of properties according to which two properties are identical just in case they are either contingently or necessarily co-exemplified (for a property like *being red* to be exemplified is just for something, say, an apple, to have that property; for two properties to be co-exemplified is for them to be jointly possessed by an object). This assumption allows the physicalist to identify the relevant property in the Knowledge Argument (*being red, being an-appearing-of-red*) with a property employed in physical theory isomorphic with it (*being brain state A*). According to the course-grained theory, if *being an-appearing-of-red* and *being brain state A* occur together, they are really just the same property. But the course-grained theory is false. *Being triangular* and *being trilateral* are different properties even though necessarily co-exemplified.

Third, when Churchland and Papineau describe Mary's new know-how, they help themselves to a number of notions that clearly seem to be dualist ones. They are listed in italics above: pre-linguistic representation, image, first-person introspection, and so forth. These dualist notions are the real meat of the physicalist rejoinder. Remove the dualist language and replace it with notions that can be captured in physicalist language, and the physicalist response becomes implausible.

CONSCIOUSNESS AND INTENTIONALITY

The third argument for property/event dualism is based on intentionality: Some (perhaps all) mental states have intentionality. No physical state has intentionality. Therefore, (at least) some mental states are not physical. As already noted, intentionality is the "of-ness" or "about-ness" of various mental states (for example, my sensation *of* a sunset, my thought *about* the weather today). Consider the following facts about intentionality:

- When one introspectively focuses one's attention on one's own conscious states, there are no sense data associated with these conscious states (e.g., a thought isn't colored); this is not so with physical states and their relations (a region of the brain in a certain state may be colored grey).
- Intentionality is completely unrestricted with regard to the kind of object it can hold as a term—anything whatever can have a mental act directed upon it, but physical relations only obtain for a narrow range of objects (e.g., magnetic fields only attract certain things).
- To grasp a mental act one must engage in an introspective act of self-awareness, but no such introspection is required to grasp a physical relation.
- For ordinary physical relations (e.g., *x* is to the left of *y*), *x* and *y* are identifiable objects irrespective of whether they have entered into that relation (if a desk is to the left of a lamp, the two objects could be arranged so as not to stand in that spatial relation to each other and, yet, the two objects would still exist); this is not so for intentional contents (e.g., one and the same belief cannot be about a frog and later

about a house—the belief is what it is, at least partly, in virtue of what the belief is *about*).
- For ordinary physical relations, each of the things standing in those relations must exist in order for the relation to obtain (e.g., x and y must exist before one can be on top of the other); but intentionality can be of nonexistent things (e.g., one can think of Zeus).

Many physicalists, especially *functionalists, try to reduce intentionality to physical causal/functional relations. Functionalists describe mental properties/states in terms of bodily inputs, behavioral outputs, and other mental state outputs. For example, a pain is whatever state is produced by pin sticks, etc., and which causes a tendency to grimace and desire pity. The state of desiring pity is, in turn, spelled out in terms of other mental states and bodily outputs. Mental properties are functional kinds. A pain is not a state with a certain intrinsic quality, namely, hurtfulness; rather, it is a function that an organism performs. Thus, a completely inanimate computer can be "conscious"—for example, "have a thought" or "be in pain"—if it performs the right outputs (produces "Ouch!") given the correct inputs (being stuck with a pin).

Dualists respond by offering thought experiments in which causal/functional relations are not the same thing as intentionality. An example of this response comes from John Searle and is known as the *Chinese Room Argument*:

> Imagine that you are locked in a room, and in this room are several baskets full of Chinese symbols. Imagine that

you (like me) do not understand a word of Chinese, but that you are given a rule book in English for manipulating the Chinese symbols. The rules specify the manipulations of symbols purely formally, in terms of their syntax, not their semantics. So the rule might say: "Take a squiggle-squiggle out of basket number one and put it next to a squoggle-squoggle sign from basket number two." Now suppose that some other Chinese symbols are passed into the room, and that you are given further rules for passing back Chinese symbols out of the room. Suppose that unknown to you the symbols passed into the room are called "questions" by the people outside the room, and the symbols you pass back out of the room are called "answers to the questions." Suppose, furthermore, that the programmers are so good at designing the programs and that you are so good at manipulating the symbols, that very soon your answers are indistinguishable from those of a native Chinese speaker. There you are locked in your room shuffling your Chinese symbols and passing out Chinese symbols in response to incoming Chinese symbols. . . . Now the point of the story is simply this: by virtue of implementing a formal computer program from the point of view of an outside observer, you behave exactly as if you understood Chinese, but all the same you don't understand a word of Chinese.[5]

The Chinese room with the person inside would simulate a computer to an outside person and represents a functionalist account of mental states like thinking and understanding meaning. For a person outside, the room receives input and gives output in a way that makes it appear that the room understands Chinese. But, of course, all the room does is imitate mental understanding—it does not possess it. Computers are just like the Chinese room. They imitate mental operations, but they do not really exemplify them. The Chinese Room Argument illustrates that functionalist accounts of mental properties and states fail to capture the role intentionality plays in our mental lives. Although all the expected inputs and outputs may be in place, functionalism doesn't account for meaning or understanding (e.g., understanding the meaning of Chinese characters), which are irreducibly mental (not physical) properties.

We have seen a number of reasons to affirm property dualism in light of the fact that consciousness and conscious states are mental and not physical. We now turn to an examination of some central arguments against property dualism.

TWO PHILOSOPHICAL OBJECTIONS TO PROPERTY DUALISM

The Problem of Causal Interaction

Physicalists claim that on a dualist construal of a human being, mind and body are so different that it seems impossible to explain how and where the two different entities interact. How could a soul, totally lacking in any physical properties, cause things to happen to the body or vice versa? How can the soul move the arm? How can a pin-stick in the finger cause pain in the soul?

This objection assumes that if we do not know *how* A

causes B, then it is not reasonable to believe *that* A causes B, especially if A and B are different. But this assumption is not a good one. We often know that one thing causes another without having any idea of how the causation takes place, especially when the two items are different. Even if one is not a theist, it is not inconceivable to believe it possible for God, if He exists, to create the world or to act in that world, even though God and the material universe are very different. Even if we grant that there is no God, if the idea of a spiritual God causing things to happen in a material world is not in itself unintelligible, then it is hard to see why a similar idea that a human soul can exercise free will and raise an arm is problematic.

Further, a magnetic field can move a tack, gravity can act on a planet millions of miles away, protons exert a repulsive force on each other, and so forth. In these examples, we know *that* one thing can (or could) causally interact with another thing, even though we may have no idea *how* such interaction takes place. Moreover, in each case the cause would seem to have a different nature from the effect—forces and fields versus solid, spatially located, particle-like entities, which doesn't seem remarkably different from a Divine Spirit interacting with a material world, or a human soul interacting with a flesh and blood body.

In the case of mind and body, we are constantly aware of causation between them. Episodes in the body or brain (being stuck with a pin, having a head injury) can produce effects in the soul (a feeling of pain, loss of memory), and the soul can cause things to happen in the body (worry can cause ulcers, one can freely and intentionally raise his arm).

We have such overwhelming evidence *that* causal interaction takes place, that there is no sufficient reason to doubt it.

Furthermore, it may even be that a *how* question regarding the interaction between mind and body cannot even arise. A question about *how* A causally interacts with B is a request for an intervening mechanism between A and B that can be described. One can ask how turning the key starts a car because there is an intermediate electrical system between the key and the car's engine that is the means by which turning the key causes the engine to start. The *how* question is a request to describe that intermediate mechanism. But the interaction between mind and body may be, and most likely is, direct and immediate. There likely *is* no intervening mechanism and thus a *how* question describing that mechanism does not even arise.

The Problem of Other Minds

The problem of other minds is this: If dualism is true, we can never know that other people have mental states because those states are private mental entities to which outsiders have no direct access. In this regard, dualism implies skepticism in two ways: It leaves us skeptical as to whether or not other minds exist in the first place, and even if they do, it leaves us skeptical as to what other persons' mental states are like. Perhaps they have completely different experiences of qualities compared with me—they sense redness and joy when I sense blueness and pain and vice versa. When my daughter was in fifth grade, she actually asked me how we knew that it wasn't the case that when her mother saw a red object, she saw it as blue but used the word "red" to talk about it, while everyone else saw it as red and used the word "red" just like mom did! If dualism is true, the objector continues, we could never know.

The dualist problem of other minds has been greatly exaggerated. For one thing, dualism does, in fact, imply the following: From what we know about a person's brain, nervous system, and behavior, we cannot logically deduce his or her mental states. But, again, far from being a vice, this implication seems to be the way things really are. That the dualist is correct here is so commonsensical, that even young children like my daughter occasionally wonder if they may sense colors in a way different from others. In general, it *is* in fact logically possible for one person to be in one kind of mental state and another person to be in a different kind of mental state even though their physical states are the same.

Second, the logical possibility just mentioned does not imply skepticism about other minds. We all know many things, for example, that we had coffee this morning, even though it is logically possible that we may be mistaken. There are many dualist views as to how we have knowledge of other minds—for example, we postulate that others are in pain when we observe them stuck with a pin and grimacing as a simple inference to the best explanation of these facts or based on an analogy with what we know we would be experiencing in a similar situation. But regardless of how we explain our knowledge of other minds, we do, in fact, have such knowledge and the mere logical possibility that we are wrong about the mental states of another is not sufficient to justify skepticism.

TWO SCIENTIFIC OBJECTIONS TO PROPERTY DUALISM

Dualism and Evolution

It is well known that one of the driving forces behind physicalism is evolutionary theory. Evolutionist Paul Churchland makes this claim:

> [T]he important point about the standard evolutionary story is that the human species and all of its features are the wholly physical outcome of a purely physical process. . . . If this is the correct account of our origins, then there seems neither need, nor room, to fit any nonphysical substances or properties into our theoretical account of ourselves. We are creatures of matter. And we should learn to live with that fact.[6]

In other words, this objection claims the following: Since humans are merely the result of an entirely physical process (the processes of evolutionary theory) working on wholly physical materials, then humans are wholly physical beings. Something does not come into existence from nothing, and if a purely physical process is applied to wholly physical materials, the result will be a wholly physical thing, even if it is a more complicated arrangement of physical materials.

Dualists could point out that this objection is clearly question-begging. To see this, note that the objection can be put into the logical form known as *modus ponens* (If P then Q; P; therefore, Q): If humans are merely the result of naturalistic, evolutionary processes, then physicalism is true.

Humans are merely the result of naturalistic, evolutionary processes. Therefore, physicalism is true.

However, the dualist could adopt the *modus tollens* form of the argument (If P, then Q; not Q; therefore, not P): If humans are merely the result of naturalistic, evolutionary processes, then physicalism is true. But physicalism is not true. Therefore, it is not the case that humans are merely the result of naturalistic, evolutionary processes. In other words, the evolutionary argument begs the question against the dualist. If the evidence for dualism is good, then the *modus tollens* form of the argument should be embraced, not the *modus ponens* form.

In fact, the existence of finite minds can be used as evidence for the existence of God. If we grant what Churchland implicitly acknowledges, namely, that there is no scientific explanation for the origin of mind, including no evolutionary explanation, then if scientific and theistic explanations are the best live alternatives, we can explain the origin of finite minds best by appealing to a Divine Mind as its most adequate cause. If we limit the alternatives to what are live options for most people in Western culture, "in the beginning" were either the particles or the Logos (Mind). It is easier to see how finite mind could come from a universe created by a Mind than it is to see how mind could come from non-rational particles.[7]

Science Makes Dualism Implausible

In one way or another, this scientific objection implies that, while possible, scientific evidence makes dualism quite unlikely. To cite one example, Nancey Murphy claims that physicalism is not primarily a philosophical thesis, but the core commitment of scientific research for which there is ample evidence. This evidence consists in the fact that "biology, neuroscience, and cognitive science have provided accounts of the dependence on physical processes of *specific* faculties once attributed to the soul."[8] Dualism cannot be *proven* false—a dualist can always appeal to correlations or functional relations between soul and brain/body—but advances in science make it a view with little justification. According to Murphy, "science has provided a massive amount of evidence suggesting that we need not postulate the existence of an entity such as a soul or mind in order to explain life and consciousness."[9]

But this is hard to substantiate. Examining the relevant arguments in this and the next chapter makes evident that science could not even formulate, much less resolve, most of the issues. For example, even if certain mental states are dependent upon specific regions of the brain (and there is evidence for dependency in the other direction as well), a dualist can explain the dependence as a form of correlation or causation, rather than as some sort of identity relation. Indeed, the central issues regarding the mind—what is a thought, feeling, or belief; what is my self identical to—are basically commonsense and philosophical issues for which scientific discoveries are largely irrelevant. Science is helpful in answering questions about what factors in the brain and body generally hinder or cause mental states to obtain, but science is largely silent about the nature of mental states themselves.

This fact is sometimes acknowledged by scientists themselves, even if begrudgingly. To illustrate, in a recent article on consciousness and neuroscience, two physicalists—Nobel prize winner Francis Crick and Christof Koch—acknowledge that one of the main attitudes among neuroscientists is that the nature of consciousness is "a philosophical problem, and so best left to philosophers."[10] Elsewhere, they claim that "scientists should concentrate on questions that can be experimentally resolved and leave metaphysical speculations to 'late-night conversations over beer.' "[11] In their scientific work, Crick and Koch choose to set aside philosophical questions about the nature of consciousness and the self, and study the neural correlates of consciousness and the causal/functional role of conscious states.

It is not science *per se*, but philosophical or methodological naturalism that is the main dualist opponent here, and dualists argue that naturalists beg important questions in their employment of science to justify physicalism. In most cases, physicalism and dualism are empirically equivalent theses (i.e., consistent with the same set of empirical observations of the brain and body) and, in fact, there is no non-question-begging theoretical virtue (e.g., simplicity, fruitfulness) that can settle the debate if it is limited to being a scientific debate. Read any book in philosophy of mind, including the earlier chapters of this book, look at the issues and arguments central to the mind/body problem, and it becomes evident that science cannot formulate, much less resolve, those issues. (The next section responds to various physicalist accounts of the mind and wades into some pretty deep philosophical waters. Thus, some readers may wish to skip this section and move on to chapter 4.)

A CRITIQUE OF PHYSICALIST ALTERNATIVES TO PROPERTY DUALISM

To understand contemporary versions of physicalism, it is important to say a word about reduction. Currently, five different types of reduction are relevant to mind/body debates:

1. **individual ontological reduction*: One object (a macro-object like a squirrel, an atom, or a person) is identified with another object or taken to be entirely composed of parts characterized by the reducing sort of entity. For example, some argue that living things are identical to, or composed entirely of, collections of physical/chemical parts, arranged in a certain way, and thus living things do not possess a soul or vital entity that accounts for their unity and status as living things. Living things are therefore *reduced* to physical or chemical components.
2. **property ontological reduction*: One property (heat) is identified with another property (mean kinetic energy).
3. **linguistic reduction*: One word or concept (*pain*) is defined as or analyzed in terms of another word or concept (*the tendency to grimace when stuck with a pin*). These kinds of reductions are definitional and may be found in the specialized vocabulary of a particular discipline.
4. **causal reduction*: The causal activity of the reduced entity is entirely explained in terms of the causal activity of the reducing entity.

5. *theoretical or explanatory reduction*: One theory or law is reduced to another by biconditional bridge principles (for example, x has heat if and only if x has mean kinetic energy). Terms in the reduced theory are connected with terms in the reducing theory by way of biconditionals that serve as the grounds for identifying the properties expressed by the former terms with those expressed by the latter. For example, if one takes color terms to be coextensional (having the same properties in common) with wavelength terms, then one can claim that colors are identical to wavelengths. Supposedly, the laws of thermodynamics can be reduced to the laws of statistical mechanics and, on that basis, heat can be identified with molecular motion.

Individual ontological reduction is affirmed by virtually all physicalists. Property reduction is affirmed by type-identity physicalists and eschewed by token physicalists and eliminativists (see below). There is a debate about whether or not functionalists accept property reduction (see below), but apart from emergent supervenient physicalists (see below), all physicalists believe that in the actual world all the properties exemplified by persons are physical properties in some sense or another. Causal reduction is hotly disputed by physicalists. Part of the debate involves the causal closure of the physical (roughly, the notion that if we trace the antecedent causes of a physical event, we will never have to leave the physical domain; there is no room, no gap, for something non-physical to cause things to happen in a chain of physical events) and the reality of so-called top-down causation. It is safe to say that, currently, most physicalists accept causal reduction. With the demise of philosophical behaviorism and positivist theories of meaning, linguistic reduction is no longer a main part of the debate. Finally, theoretical reduction is the main type of reduction employed in classifications of physicalism and, unless otherwise indicated, descriptions of reductive vs. nonreductive physicalism should be understood to employ it.

Currently, the main version of reductive physicalism is *type-identity physicalism. Type-identity physicalists accept explanatory reduction and, on that basis, property ontological reduction. On this view, mental properties/types are identical to physical properties/types. Moreover, identity statements asserting the relevant identities are construed as contingent identity statements employing different yet co-referring expressions. For example, the statement "Red is identical to wavelength X" is contingently true (while true, it could have been false, unlike "2+2=4," which is a *necessary* truth), and the terms "red" and "wavelength X" both refer to the same thing (namely, a specific wavelength), even though the terms do not have the same dictionary definition. Likewise, "Painfulness is identical to a type of C-fiber firing pattern" is allegedly a contingent identity statement. The truth of these identity statements is an empirical discovery, and the statements are theoretical identities.

Two main objections seem decisive against type-identity physicalism. First, it is just obvious that mental and physical properties are different from each other, and physicalists have not met the burden of proof required to overturn these deeply ingrained intuitions. Earlier in this chapter we looked at some of these differences.

Physicalists respond that in other cases of identity (heat is mean kinetic energy), our intuitions about non-identity turned out to be wrong, and the same is true in the case of mental properties. But for two reasons, this response fails. For one thing, the other cases of alleged property identities are most likely cases of correlation of properties.

Second, we can easily explain why our intuitions were mistaken (granting that they were mistaken for the sake of argument) in the other cases, but a similar insight does not appear in the case of mental properties.[12] Since there is a distinction between what heat is (mean kinetic energy) and how it appears to us (as being warm), our intuitions about non-identity confused appearance with reality. But since mental properties such as painfulness are identical to the way they appear (a pain just is the way it seems to us, namely, as a hurtful state), no such source of confusion is available. Thus, intuitions about their non-identity with physical properties remain justified.

The second difficulty with type-identity theory is called the multiple-realization problem, though a more accurate label would be the multiple-exemplification problem, since according to dualists, mental properties are exemplified and not realized (see below). In any case, it seems obvious that humans, dogs, Vulcans, and a potentially infinite number of organisms with different physical-type states can all be in pain and thus the mental kind, being painful, is not identical to a physical kind.

Largely in response to this last problem, a version of

(allegedly) nonreductive physicalism—functionalism—has become the prominent current version of physicalism. Functionalists employ a topic-neutral description of mental properties/states in terms of bodily inputs, behavioral outputs, and other mental state outputs. "Topic neutral" means a characterization of a mental state in terms that are neutral as to whether the state turns out to be physical or mental. Such a characterization depicts a mental state in terms of its functional role in behavior, not in terms of its intrinsic attributes.

For example, a pain is whatever state is produced by pin sticks, etc., and which causes a tendency to grimace and desire pity. The state of desiring pity is, in turn, spelled out in terms of other mental states and bodily outputs. Mental properties are functional kinds. Machine functionalists characterize the various relations that constitute a functional state in terms of abstract computational, logical relations; and causal-role functionalists spell them out in terms of causal relations. Either way, a mental property such as painfulness turns out to be the second-order property (a second-order property is a property of a property, for example, being colored is a property of being red) of having a property that plays the relevant functional role R. In this way, "mental properties" are treated very much like computer software. Type-identity physicalism is a hardware view; functionalism is a software view.

Functionalism may also be characterized in terms of supervenience (see below, and glossary) and realization (see below). There are different views of supervenience, but here is a standard formulation of (strong) property supervenience:

(SS) Mental properties supervene on physical properties, in that necessarily any two things (in the same possible world or in different possible worlds with the same laws of nature) indiscernible in all physical properties are indiscernible in mental properties.

The basic idea here is that the physical properties of an object are sufficient to fix its "mental" properties. The object cannot change in its "mental" properties without first changing in its physical properties.

The *realization* relation between mental property M and physical property P may be characterized in this way:

(Rz) For some entity *x*, *P* realizes *M* just in case *x* is *M* in virtue of *x* being *P*.

An example would be the way a diskette "realizes" being a certain sort of software. For the functionalist, P realizes M just in case P has the property of satisfying functional role R in x. In this way, functionalism is a form of nonemergent, *structural supervenience such that mental properties supervene on physical properties and the latter realize the former. More specifically, if P realizes M, then M supervenes on P.

There are at least three serious difficulties with functionalism in its various formulations. First, there are problems regarding absent or inverted *qualia. A quale (pronounced "kwa-lee"; plural, qualia) is a specific sort of intrinsically characterized mental state, such as seeing red, having a sour taste, feeling a pain. If a Vulcan realized the correct functional role for pain but exemplified the causal intermediary of being appeared to redly while feeling no hurtfulness at all, the functionalist would say that the Vulcan is in pain. But it seems obvious that the Vulcan is not in pain but, rather, is experiencing the mental state of an appearing-of-red, and this supports property dualism. Qualia arguments turn on the observation that mental kinds are essentially characterized by their intrinsic properties (their inherent feel) and contingently characterized by their extrinsic functional relations. Thus, property dualism correctly captures the essence of mental properties, and functionalism fails on this score.

Second, for two reasons, functionalism fails to account for first-person knowledge of one's own mental states by introspection. For one thing, on a functionalist interpretation, what makes a specific pain event a pain has nothing whatever to do with its intrinsic features. What makes a specific mental event the kind of mental event it is, is entirely a matter of its extrinsic relations to inputs from the environment and bodily and other mental outputs. The same brain event that realizes pain in one possible world could realize any other mental state in a different world with a different functional context. Thus, there is no way for a subject to know what mental state he is having by being conscious of it. In fact, since bodily outputs are essential for characterizing the mental kind, one would have to wait until he observed his own behavior to see what mental state he was in!

Moreover, since one factor that constitutes a given mental state is its relationship to a further mental output, it is hard for a functionalist to avoid *mental holism—roughly, the notion that a given mental state gets its identity from its entire set of relations to all the other mental states in one's entire psychology. Even if the functionalist can delimit a subset of one's psychology to serve as the relevant set (and this is doubtful), the problem still remains that one could not know what a given mental state is by attending to it without running through the entire set of internally related psychological states that constitute it. Experientially, this is not how people know their own mental states, and, in any case, the functionalist notion that a mental state is constituted by both bodily outputs and other mental states makes problematic introspective knowledge of one's thoughts, sensations, etc. There can be no physics of money, and no amount of inspection of the intrinsic physical properties of a dollar bill will give any clue that it is money. Assignments of economic characteristics (e.g., the monetary value) are arbitrary as far as the intrinsic features of the realizers (e.g., the paper, the ink, the graphic design) are concerned. The same may be said about awareness of and characteristic assignments for the realizers of functionally characterized mental kinds.

Finally, there is no clear sense as to what the realization relation is that is available to a strict physicalist. Realization is not a relation that figures into chemistry and physics. Further, when physicalists characterize the realization relation or realizers that stand in it in terms of clear cases, they select artifacts and appeal to mental notions such as intentions, values, goals, and agent production in their characterization (for example, a certain piece of paper exemplifies the property of being rectangular, but it merely realizes the "property" of being a dollar in that the piece of paper plays a certain role in the economic lives of people, and the economic realm involves people's goals [to buy a car], values ["we should be paid for the time we work"], and so forth).[13] If the objective is to characterize either realization itself or what it is to be a realizer in non-mental terms, this will hardly do. Moreover, dualists claim that mental properties are not realized; rather, they are exemplified (which allows for the fact that mental states are compatible with a number of different physical states) and thereby constitute that which grounds the unity of mental kinds.

The next view, *token physicalism, is a hard view to classify. Fundamentally, it amounts to the claim that, even though there is no smooth property identity for mental types, every token (that is, particular) mental event is identical to a particular physical event. For example, there is not a single type of physical property that Vulcans, dogs, humans, etc., must have to count as being in pain. But, on this view, every time any organism is in pain, that particular pain event will be identical to some brain event or other. Beyond that, things are not so clear and it is beyond the scope of this chapter to probe this viewpoint more deeply. However, it is safe to say that for most physicalists, token physicalism is not a distinct viewpoint; rather, it is part of the specification of a full-blown physicalist functionalism such that mental properties are functional types and particular physical events are the token realizers of those types. So understood, the objections raised against functionalism also apply to token physicalism.

Finally, *eliminative materialism is roughly the claim that mental terms get their meaning from their role in *folk psychology (roughly, a commonsense theory designed to explain the behaviors of others [grimacing] by attributing mental states [pain] to them), and, like the debunked Phlogiston theory, folk psychology will eventually be replaced with a neurophysiological theory. On this view, the various mental terms of folk psychology fail to refer to anything and should be eliminated. Some eliminative materialists apply the view to all mental states (including sensations such as pain) while others limit it to propositional states such as beliefs and thoughts.

Eliminative materialism has not garnered widespread acceptance. First, dualism is not primarily a theory, much less a replaceable one; rather, it is a descriptive report of the mental self and its states with which one is acquainted through introspection. Second, it simply seems implausible to say that no one has ever actually had a sensation or belief. Third, some have argued that in effect eliminative materialism is self-refuting in that it advocates the belief that there are no such things as beliefs. Some eliminative materialists have responded that self-refutation can be avoided because, while their view does in fact reject the existence of beliefs, it does allow for a physical replacement that plays the same role as beliefs and this replacement is what the theory advocates. But many critics remain skeptical of this response on the grounds that if an entity is found that actually plays the same role as a belief, it will be a belief by another name. If it plays a different role, then self-refutation may be avoided only at the expense of proffering an inadequate revisionism.

Finally, a word should be said about supervenient physicalism. *Emergent supervenience is the view that mental properties are distinctively new kinds of properties that in no way characterize the subvenient physical base on which they depend. So understood, it is actually a form of property dualism. Structural supervenience is the view that mental properties are structural properties entirely constituted by the properties, relations, parts, and events at the subvenient level. Functionalism is currently the most popular version of structural supervenience.

Taken by itself, supervenient physicalism is not a distinct viewpoint. Alone, it fails to capture property dependence (the dependence of mental properties on physical ones) and, instead, only expresses covariance between mental and physical properties. So understood, it is consistent with substance dualism, type physicalism, and epiphenomenalism. In fact, it allows for cases where A supervenes on B, yet B is in some sense dependent on A. Personhood supervenes on being human, but arguably, this is a genus/species relation in which species are ontologically dependent on their genera for their existence and identity.

In order for supervenience to express the dependence of mental properties on physical properties and, thus, to be adequate for at least minimal physicalism, it must be supplemented with two further principles:

(1) The *anti-Cartesian principle: There can be no purely mental beings (e.g., substantial human souls, God, angels) because nothing can have a mental property without having a physical property as well.
(2) *Mind-body dependence: What mental properties an entity has depend on and are determined by its physical properties.

By employing arguments in this and the next chapter, property and substance dualists will reject (1) and (2). Moreover, since it is strictly a metaphysical thesis, there is no scientific evidence that could justify (1), so the authority of science cannot be claimed on its behalf. Regarding (2), there is scientific evidence for the dependency it expresses. But there is also scientific evidence that mental states causally affect brain states and, in any case, substance dualist arguments presented in the next chapter (e.g., the modal argument and the argument from libertarian freedom) provide counterexamples to (2).

CHAPTER IN REVIEW

In chapter 3 we saw that consciousness is a mystery, but one best explained by God's existence. While finite minds make sense in a universe created by a Divine Mind, they are exceedingly difficult to account for in a naturalistic universe. Physicalist approaches to explaining mental properties ultimately fall short because mental properties possess unique characteristics that cannot be reduced to physical states, events, and properties.

- **A Case for Property Dualism**
- Arguments for property dualism show that mental states are in no sense physical since they possess five features not owned by physical states:
  - A raw qualitative feel or a "what-it-is-like" to have a mental state.

- ☆ Intentionality—*of*-ness or *about*-ness—directed toward an object.
- ☆ Are inner, private, and immediate to the subject having them.
- ☆ Require a subjective ontology.
- ☆ Fail to have crucial features that characterize physical states and, in general, cannot be described using physical language.
- The Argument from Introspection: Attending to one's own mental life reveals that mental states have a variety of features that physical states do not have, such as those listed above. Therefore, mental states are not identical to physical states.
- The Knowledge Argument: The Mary thought experiment demonstrates that there is knowledge of facts that cannot be physical facts, but must be mental facts.
- The Argument from Intentionality: This argument demonstrates that some (perhaps all) mental states have intentionality. No physical state has intentionality. Therefore, (at least) some mental states are not physical.

- **Philosophical Objections to Property Dualism**
- The Problem of Causal Interaction: According to dualism, mind and body are so different that it seems impossible to explain how and where the two different entities interact.
- ☆ Reply: This objection assumes that if we do not know *how* A causes B, then it is not reasonable to believe

that A causes B, especially if A and B are different. But this assumption is problematic.

- The Problem of Other Minds: If dualism is true, we can never know that other people have mental states because those states are private mental entities to which outsiders have no direct access.
- ☆ Reply: First, this implication seems to be the way things really are (we in fact don't have access to others' mental states), which counts in favor of dualism. Second, the mere logical possibility that we are wrong about the mental states of another is not sufficient to justify skepticism.

- **Two Scientific Objections to Dualism**
- Dualism and Evolution: Since humans are merely the result of an entirely physical process working on wholly physical materials, then humans are wholly physical beings.
- ☆ Reply: This argument begs the question against the dualist since it begins by assuming humans are wholly physical beings.
- Science Makes Dualism Implausible: Advances in science make dualism a view with little justification.
- ☆ Reply: This argument is hard to substantiate. Moreover, the most significant issues in this debate are not scientific, but philosophical and commonsensical.

- **A Critique of Physicalist Alternatives to Property Dualism**
- Type-Identity Physicalism: The view that mental

properties/types are identical to physical properties/types.

☆ Critique: First, it is obvious that mental and physical properties are not identical. Second, it seem obvious that a potentially infinite number of organisms with different physical-type states can all be in pain and thus the mental kind, being painful, is not identical to a physical kind.

- Functionalism: The view that reduces mental properties/states to bodily inputs, behavioral outputs, and other mental state outputs.

☆ Critique: First, the inverted qualia argument shows that mental kinds are essentially characterized by their intrinsic properties and only contingently characterized by their extrinsic functional relations. Second, functionalism fails to account for first-person knowledge of one's own mental states by introspection. Finally, there is no clear sense as to what the realization relationship is according to strict physicalism.

- Token Physicalism: Fundamentally, the claim that every token (that is, particular) mental event is identical to a particular physical event.

☆ Critique: The criticisms of functionalism apply to token physicalism.

- Eliminative Materialism: The view that mental terms get their meaning from their role in folk psychology, and will eventually be replaced with some neurophysiological theory.

☆ Critique: First, that no one has ever actually had a

sensation or belief is implausible. Second, eliminative materialism is self-refuting in that it advocates the belief that there are no such things as beliefs.

KEY VOCABULARY

Anti-Cartesian principle: There can be no purely mental beings because nothing can have a mental property without having a physical property as well.

Causal reduction: The causal activity of the reduced entity is entirely explained in terms of the causal activity of the reducing entity.

Consciousness: Broadly, what you are aware of when you engage in first-person introspection.

Eliminative materialism: Mental terms get their meaning from their role in folk psychology (see definition below), and will eventually be replaced with some neurophysiological theory.

Emergent supervenience: The view that mental properties are distinctively new kinds of properties that in no way characterize the subvenient physical base on which they depend.

Epiphenomenon: Something that is caused to exist by something else but that itself has no ability to cause anything.

Epiphenomenalism: The mind is a by-product of the brain, which causes nothing; the mind merely "rides" on top of the events in the brain.

Folk psychology: A commonsense theory designed to explain the behaviors of others by attributing mental states to them.

Functionalism: The physicalist view that reduces mental properties/states to bodily inputs, behavioral outputs, and other mental state outputs.

Individual ontological reduction: One object (a macro-object like a dog, a molecule, or a person) is identified with another object or taken to be entirely composed of parts characterized by the reducing sort of entity.

Intentionality: The "of-ness" or "about-ness" of various mental states.

Know-how: The ability to do certain things.

Knowledge by acquaintance: Knowledge of a thing when one is directly aware of that thing.

Linguistic reduction: One word or concept (*pain*) is defined as or analyzed in terms of another word or concept (*the tendency to grimace when stuck with a pin*). These kinds of reductions are definitional.

Mental holism: The notion that a given mental state gets its identity from its entire set of relations to all the other mental states in one's total psychology.

Mind-body dependence: What mental properties an entity has depend on and are determined by its physical properties.

Property dualism: A human being is one material substance that has both physical *and* mental properties, with the mental properties arising from the brain.

Property ontological reduction: One property (heat) is identified with another property (mean kinetic energy).

Propositional knowledge: Knowledge that a proposition is true.

Qualia: A quale (plural, qualia) is a specific sort of intrinsically characterized mental state, such as seeing red, having a sour taste, feeling a pain.

Self-presenting property: A property, such as *being appeared to redly* (that is, experiencing an appearance of the color red), that presents both its intentional object (e.g., a red apple) and itself (the redness) to the subject experiencing it.

Supervenience: A relationship of dependence between properties such that one level of the properties correlates to conditions at a different level. For example, when water molecules come together, the property of wetness supervenes upon them. In mind/body discussions, some philosophers (such as certain property dualists) hold that mental events supervene upon (or emerge from) brain events.

Structural supervenience: The view that mental properties are structural properties entirely constituted by the properties, relations, parts, and events at the subvenient level.

Theoretical or explanatory reduction: One theory or law is reduced to another by biconditional bridge principles (for example, x has heat if and only if x has mean kinetic energy).

Token physicalism: Fundamentally, the claim that every token (that is, particular) mental event is identical to a particular physical event.

Type-Identity physicalism: The view that mental properties/types are identical to physical properties/types.

NOTES

1. Colin McGinn, *The Mysterious Flame* (New York: Basic Books, 1999), 13–14.
2. Phillip Johnson, *The Right Questions: Truth, Meaning & Public Debate* (Downers Grove, IL: IVP, 2002), 63.
3. Thomas Nagel, "What Is It Like to Be a Bat?" *The Philosophical Review* 83 (1974): 435–50; Frank Jackson, "Epiphenomenal Qualia," *Philosophical Quarterly* 32 (1982): 127–36; Saul Kripke, "Naming and Necessity," in *Semantics of Natural Languages*, eds. Donald Davidson and Gilbert Harman (Dordrecht: D. Reidel, 1972), 253–355. Subsequently, Jackson has raised doubts about the Knowledge Argument. See his "What Mary Didn't Know," *Journal of Philosophy* 83 (1986): 291–95.
4. David Papineau, *Philosophical Naturalism* (Cambridge, MA: Blackwell, 1993), 103–14; Paul M. Churchland, *Matter and Consciousness* (Cambridge, MA: MIT Press, 1984), 33–34.
5. John Searle, *Minds, Brains, and Science* (Cambridge, MA: Harvard University Press, 1984), 32–33. Cf. John Searle, "Minds, Brains, and Programs," *The Behavioral and Brain Sciences* 3 (1980): 417–24.
6. Churchland, *Matter and Consciousness*, 21.
7. For an example of this argument, see J. P. Moreland, *Consciousness and the Existence of God* (London: Routledge, 2008).
8. Nancey Murphy, "Human Nature: Historical, Scientific, and Religious Issues," in *Whatever Happened to the Soul?* eds. Warren S. Brown, Nancey Murphy, and H. Newton Malony (Minneapolis: Fortress Press, 1998), 17. Cf. pp. 13, 27, 139–43.
9. Ibid., 18.
10. Francis Crick and Christof Koch, "Consciousness and Neuroscience," *Cerebral Cortex* 8 (1998): 97–107.
11. Cf. John Horgan, "Can Science Explain Consciousness?" *Scientific American*, July 1994, 91.
12. Saul Kripke, *Naming and Necessity* (Cambridge, MA: Harvard University Press, 1972), 148–55.
13. Cf. David Papineau, "Arguments for Supervenience and Physical Realization," in *Supervenience: New Essays*, eds. Elias E. Savellos and Umit D. Yalcin (Cambridge: Cambridge University Press, 1995), 226–43.

Chapter Four: THE REALITY OF THE SOUL

In chapter 3, I defended property dualism and concluded that consciousness is, indeed, non-physical. In this chapter, I will argue for substance dualism, the view that the owner of consciousness—the soul or self—is immaterial. Substance dualists are also property dualists, because substance dualists believe that both the ego and consciousness itself are immaterial. But one can be a mere property dualist without being a substance dualist if one accepts the immateriality of consciousness but holds that its owner is the body or, more likely, the brain. In contrast to mere property dualism, substance dualists hold that the brain is a physical thing that has physical properties, and the mind or soul is a mental substance that has mental properties. When I am in pain, the brain has certain physical properties (electrical, chemical), and the soul or self has certain mental properties (the conscious awareness of pain). The soul is the possessor of its experiences. It stands behind and above them and remains the same throughout my life. The soul and the brain can interact with each other, but they are different objects with different properties.

Currently, there are three main forms of substance dualism. First, there is *Cartesian dualism, according to which the mind is a substance with the ultimate capacities for consciousness, and it is connected to its body by way of a causal relation (that is, my body is mine because I can cause things to happen to it and vice versa).[1] Second, there is *Thomistic substance dualism, one important version of which takes the soul to be broader than the mind in containing not merely the capacities for consciousness, but also those which ground biological life and functioning. On this view, the (human) soul diffuses, informs (gives form to), unifies, animates, and makes human the body. The body is not a physical substance, but rather, an ensouled physical structure such that if it loses the soul, it is no longer a human body in a strict, philosophical sense.[2] Instead, it becomes a corpse. According to the third form, *emergent dualism, a substantial, spatially extended, immaterial self emerges from the functioning of the brain and nervous system, but once it emerges, it exercises its own causal powers and continues to be sustained by God after death.[3] As interesting as this intramural debate is, for our purposes we will set it aside and simply argue for that which all three positions hold in common: the self or ego is an immaterial substance that bears consciousness.

A CASE FOR SUBSTANCE DUALISM AND THE IMMATERIAL NATURE OF THE SELF

At least five arguments have been offered in the recent literature for some form of substance dualism.

Our Basic Awareness of the Self

Stewart Goetz has advanced the following type of argument for the nonphysical nature of the self, which I have modified:[4]

(1) I am essentially an indivisible, simple spiritual substance.
(2) Any physical body is essentially a divisible or complex entity (any physical body has spatial extension or separable parts).
(3) The law of identity pertains (if x is identical to y, then whatever is true of x is true of y, and vice versa).
(4) Therefore, I am not identical with my (or any) physical body.
(5) If I am not identical with a physical body, then I am a soul.
(6) Therefore, I am a soul.

Premise (2) is pretty obvious, and (5) is commonsensical. The body and brain are complex material objects made of billions of parts—atoms, molecules, etc. Premise (3) is the law of identity I introduced in chapter 1. Regarding premise (1), we know it is true by introspection. When we enter most deeply into ourselves, we become aware of a very basic fact presented to us: We are aware of our own self (ego, I, center of consciousness) as being distinct from our bodies and from any particular mental experience we have, and as being an uncomposed, spatially unextended, simple center of consciousness. In short, we are just aware of ourselves as simple, conscious things. This fundamental awareness is what grounds my *properly basic belief (a rational belief that is not inferred from other beliefs) that I am a simple center of consciousness. On the basis of this awareness, and premises (2) and (3), I know that I am not identical to my body or my conscious states; rather, I am the immaterial self that *has* a body and a conscious mental life.

An experiment may help convince you of this. Right now I am looking at a chair in my office. As I walk toward the chair, I experience a series of what are called phenomenological objects or chair representations. That is, I have several different chair experiences that replace one another in rapid succession. As I approach the chair, my chair sensations vary. If I pay attention, I am also aware of two more things. First, I do not simply experience a series of sense-images of a chair. Rather, through self-awareness, I also experience the fact that it is I myself who has each chair experience. Each chair sensation produced at each angle of perspective has a perceiver who is I. An "I" accompanies each sense experience to produce a series of awarenesses—"I am experiencing a chair sense-image now."

I am also aware of the basic fact that the self that is currently having a fairly large chair experience (as my eyes come to within twelve inches of the chair) is the very same self as the one who had all of the other chair experiences preceding this current one. Through self-awareness, I am aware of the fact that I am an enduring self who was and is (and will be) present as the owner of all the experiences in the series.

These two facts—I am the owner of my experiences, and I am an enduring self—show that I am not identical to my experiences. Rather, I am the conscious thing that has them. I am also aware of myself as a simple, uncomposed, and spatially unextended center of consciousness. In short, I am a mental substance. Moreover, I am "fully present" throughout my body, though not identical to any part of it; if my arm is cut off, I do not become four-fifths of a self. My body and brain are divisible and can be present in percentages (there could be 80 percent of a brain present after an operation). But *I* am an all-or-nothing kind of thing. I am not divisible; I cannot be present in percentages.

UNITY AND THE FIRST-PERSON PERSPECTIVE

Consider the following argument:

(1) If I were a physical object (e.g., a brain or body), then a third-person physical description would capture all the facts that are true of me.
(2) But a third-person physical description does not capture all the facts that are true of me.
(3) Therefore, I am not a physical object.
(4) I am either a physical object or a soul.
(5) Therefore, I am a soul.

A complete physical description of the world would be one in which everything would be exhaustively described from a third-person point of view in terms of objects, properties, processes, and their spatiotemporal locations. For example, a description of an apple in a room would go something like this: "There exists an object three feet from the south wall and two feet from the east wall, and that object has the property of being red, round, of weighing 3.5 ounces," and so on.

The *first-person point of view is the vantage point that I use to describe the world from my own perspective. Expressions of a first-person point of view utilize what are called *indexicals*—words like "I," "here," "now," "there," "then." Here and now are where and when I am; there and then are where and when I am not. Indexicals refer to me. "I" is the most basic indexical, and it refers to my self that I know by acquaintance with my own acts of self-awareness. I am immediately aware of my own self and I know to whom "I" refers when I use it: it refers to me as the conscious owner of my body and mental states.

According to a widely accepted form of physicalism, there are no irreducible, privileged, first-person perspectives. Everything can be exhaustively described in an object language from a third-person perspective. A physicalist description of me would say, "There exists a body at a certain location that is five feet eight inches tall, weighs 160 pounds," and so forth. The property dualist would add a description of the properties possessed by that body, such as "The body is feeling pain," or "It is thinking about lunch."

But no amount of third-person descriptions captures my own subjective, first-person acquaintance of my own self in acts of self-awareness. In fact, for any third-person description of me, it would always be an open question as to whether the person described in third-person terms was the same person as I am. I do not know myself *because* I know some third-person description of a set of mental and physical properties that apply to me ("So, the body is five-feet-eight-inches, 160 pounds, and is thinking about lunch? I think that's me."). Instead I know myself as a self immediately through being acquainted with my own self in an act of self-awareness. I can express that self-awareness by using the term *I*.

I refers to my own substantial soul. It does not refer to any mental property or bundle of mental properties I am having, nor does it refer to any body described from a third-person perspective. *I* is a term that refers to something that exists, and *I* does not refer to any object or set of properties described from a third-person point of view. Rather, *I* refers to my own self with which I am directly acquainted and which, through acts of self-awareness, I know to be the substantial, uncomposed possessor of my mental states and my body.

A related argument has been offered by William Hasker:

(1) If I am a physical object (e.g., a brain or a body), I do not have a unified visual field.
(2) I do have a unified visual field.
(3) Therefore, I am not a physical object.
(4) I am either a physical object or a soul.
(5) Therefore, I am a soul.

To grasp the argument, consider one's awareness of a complex fact, say, one's own visual field consisting of awareness of several objects at once, including a number of different surface areas of each object. Now, one may claim that such a unified awareness of one's visual field consists in the fact that there are a number of different physical parts, each of which is aware only of part of and not the whole of the complex fact. Indeed, this is exactly what physicalists say. We now know that when one looks at an object, different regions of the brain process different electrical signals that are associated with different aspects of the object (e.g., its color, shape, size, texture, location).[5] However, this claim will not work, because it cannot account for the fact that there is a single, unitary awareness of the entire visual field.[6] There is no region in the brain that "puts the object back together into a unified whole." Apart from a unifying consciousness, our daily lives would consist of a series of unconnected perceptions and images completely lacking in coherence. Only a single, uncomposed mental substance can account for the unity of one's visual field or, indeed, the unity of consciousness in general.

THE MODAL ARGUMENT

The core of the modal argument for the soul is fairly simple: I am possibly disembodied (I could survive without my brain or body); my brain or body are not possibly disembodied (they could not survive without being physical); so I am not my brain or body. I am either a soul or a brain or a body, so I am a soul. Let's elaborate on the argument.

Thought experiments such as this one have rightly been central to debates about personal identity. For example, we are often invited to consider a situation in which two people switch bodies, brains, or personality traits, or in which a person exists disembodied. In these thought experiments, someone argues in the following way: Because a certain state of affairs S (e.g., Smith existing disembodied) is conceivable, this provides justification for thinking that S is metaphysically possible. Now if S is possible, then certain implications follow about what is/is not essential to personal identity (e.g., Smith is not essentially a body).

We all use conceiving as a test for possibility/impossibility throughout our lives. I know that life on other planets is possible (even if I think it is highly unlikely or downright false) because I can conceive it to be so. I am aware of what it is to be living and to be on earth and I conceive no necessary connections between these two properties (surely life *could* exist on other planets besides this one). On the other hand, I know square circles are impossible because they are inconceivable given my knowledge of being square and being circular. To be sure, judgments that a state of affairs is possible/impossible grounded in conceivability are not infallible. They can be wrong. Still, they provide strong evidence for genuine possibility/impossibility. In light of this, I offer the following criterion:

> For any entities x and y, if I have good grounds for believing I can conceive of x existing without y (e.g., a dog without being colored brown) or vice versa, then I have good grounds for believing x (being brown) is not essential or identical to y (being a dog) or vice versa.

Let us apply these insights about conceivability and possibility to the modal argument for substance dualism. We can expand the argument outlined above as follows:[7]

(1) The law of identity: If x is identical to y, then whatever is true of x is true of y and vice versa.
(2) I can strongly conceive of myself as existing disembodied. (For example, I have no difficulty believing that out-of-body near-death experiences are possible; that is, they *could* be true.)

(3) If I can strongly conceive of some state of affairs S (e.g., my disembodied existence) that S possibly obtains, then I have good grounds for believing that S is possible.
(4) Therefore, I have good grounds for believing of myself that it is possible for me to exist and be disembodied.
(5) If some entity x (for example, my self) is such that it is possible for x to exist without y (for example, my brain or body), then (i) x (my self) is not identical to y (my brain or body) and (ii) y (my brain or body) is not essential to x (me).
(6) My body (or brain) is not such that it is possible to exist disembodied, i.e., my body (or brain) is essentially physical.
(7) Therefore, I have good grounds for believing of myself that I am not identical to my body (or brain) and that my physical body is not essential to me.

A parallel argument can be advanced in which the notions of a body and disembodiment are replaced with the notions of physical objects. So understood, the argument would imply the conclusion that I have good grounds for thinking that I am not identical to a physical object, nor is any physical object essential to me. A parallel argument can also be developed to show that possessing the ultimate capacities of sensation, thought, belief, desire, and volition *are* essential to me, implying that I am a substantial soul or mind and I could not exist without the ultimate capacities of consciousness.

I cannot undertake a full defense of the argument here, but it would be useful to say a bit more regarding (2). There are a number of things about ourselves and our bodies of which we are aware that ground the conceivability expressed in (2). I am aware that I am unextended (I am "fully present" at each location in my body, as Augustine claimed; I occupy my body as God occupies space by being fully present throughout it); I am not a complex aggregate made of substantial parts, nor am I the sort of thing that can be composed of physical parts; rather, I am a basic unity of inseparable faculties (of mind, volitions, emotion, etc.) that sustains absolute sameness through change; and I am not capable of gradation (I cannot become two-thirds of a person).[8]

In near-death experiences, people report themselves to have been disembodied. They are not aware of having bodies in any sense. Rather, they are aware of themselves as unified egos that have sensations, thoughts, and so forth. Moreover, Christians who understand the biblical teaching that God and angels are bodiless spirits also understand by direct introspection that they are like God and angels in the sense that they are spirits with some of the same sorts of powers God and angels have (e.g., to think rationally, to make moral judgments) but that they also have bodies. In addition, the New Testament teaching on the intermediate state is intelligible in light of what they know about themselves and it implies that we will and, therefore, *can* exist temporarily without our bodies. In 2 Corinthians 12:1–4, Paul asserts that he may actually have been disembodied. Surely part of the grounds for Paul's willingness to consider this a real possibility was his own awareness of his nature through introspection, his recognition of his similarity to God and angels in this respect, and his knowledge of biblical teaching. All of the factors discussed above imply that people can conceive of themselves as existing in a disembodied state, and they provide grounds for thinking that this is a real possibility (even if it turns out to be false—though, of course, I do not think it is). If such disembodiment is even possible, then one cannot be one's body, nor is one's body essential to him.

FREE WILL, MORALITY, RESPONSIBILITY, AND PUNISHMENT

Consider the following argument:

(1) If I am a physical object (e.g., a brain or a body), then I do not have free will.
(2) But I do have free will.
(3) Therefore, I am not a physical object.
(4) I am either a physical object or a soul.
(5) Therefore, I am a soul.

When I use the term *free will*, I mean what is called libertarian freedom. I can literally choose to act or refrain from acting. No circumstances exist that are sufficient to determine my choice. My choice is up to me. I act as an agent who is the ultimate originator of my own actions. Moreover, my reasons for acting do not partially or fully cause my actions; I myself bring about my actions. Rather, my reasons are the teleological goals or purposes for the sake of which I act. If I get a drink because I am thirsty, the desire to satisfy my thirst is the end for which I myself act freely. I raise my arm *in order to* vote.

If physicalism is true, then human free will does not exist. Instead, determinism is true.[9] If I am just a physical system, there is nothing in me that has the capacity to freely choose to do something. Material systems, at least large-scale ones, change over time in deterministic fashion according to the initial conditions of the system and the laws of chemistry and physics. A pot of water will reach a certain temperature at a given time in a way determined by the amount of water, the input of heat, and the laws of heat transfer.

Now, when it comes to morality, it is hard to make sense of moral obligation and responsibility if determinism is true. They seem to presuppose freedom of the will. If I "ought" to do something, it seems to be necessary to suppose that I *can* do it, that I could have done otherwise, and that I am in control of my actions. No one would say that I ought to jump to the top of a fifty-floor building and save a baby, or that I ought to stop the American Civil War in this present year, because I do not have the ability to do either. If physicalism is true, I do not have any genuine ability to choose my actions. Further, free acts seem to be teleological. We act for the sake of goals or ends. If physicalism (or mere property dualism) is true, there is no genuine teleology and, thus, no libertarian free acts.

It is safe to say that physicalism requires a radical revision of our commonsense notions of freedom, moral obligation, responsibility, and punishment. On the other hand, if these commonsense notions are true, physicalism is false.

The same problem besets mere property dualism. There are two ways for property dualists to handle human actions. First, some property dualists are epiphenomenalists (a term we encountered in chapter 3). A person is a living physical body having a mind, the mind consisting, however, of nothing but a more or less continuous series of conscious or unconscious states and events that are the effects but never the causes of bodily activity. Put another way, when matter reaches a certain organizational complexity and structure, as is the case with the human brain, then matter produces mental states like fire produces smoke, or the structure of hydrogen and oxygen in water produces wetness. The mind is to the body as smoke is to fire. Smoke is different from fire (to keep the analogy going, the physicalist would identify the smoke with the fire or the functioning of the fire), but fire causes smoke, not vice versa. The mind is a by-product of the brain that causes nothing; the mind merely "rides" on top of the events in the brain. Hence, epiphenomenalists reject free will since they deny that mental states cause anything. A second way that property dualists handle human action is through a notion called **event-event causation.*[10] To understand event-event causation, consider a brick that breaks a glass. The cause in this case is not the brick itself (which is a substance), but an event, namely, the brick's being in a certain state—a state of motion. And this event (the brick's being in a state of motion) was caused by a prior event, and so on. The effect is another event, namely, the glass's being in a certain state—breaking. Thus, one event—the moving of a brick—causes another event to occur—the breaking of the glass. Further, according to event-event causation, whenever one event causes another, there will be some deterministic or probabilistic law of nature that relates the two events. The first event, combined with the laws of nature, is sufficient to determine or fix the chances for the occurrence of the second event.

Agent action, on the other hand, is an important part of an adequate libertarian account of freedom of the will. One example of agent action is this typical case: my raising my arm. When I raise my arm, I, as a substance, simply act by spontaneously exercising my active powers. *I* raise my arm; I freely and spontaneously exercise the powers within my substantial soul and simply act. No set of conditions exists within me that is sufficient to determine that I raise my arm. Moreover, this substantial agent is characterized by the power of active freedom, conscious awareness, the ability to think, form goals and plans, to act teleologically (for the sake of goals), and so forth. Such an agent is an immaterial substance and not a physical object, which lacks these abilities. Thus, libertarian freedom is best explained by a substance dualism that involves agent action, and not by physicalism or mere property dualism that lacks such action.

Unfortunately for property dualists, event-event causation is deterministic. Why? For one thing, there is no room for an agent, an ego, an "I" to intervene and contribute to one's actions. On this view, I do not produce the action of raising my arm; rather, a state of desiring to raise an arm is sufficient to produce the effect. There is no room for my own self, as opposed to the mental states within me, to act.

Moreover, all the mental states within me (my states of desiring, willing, hoping) are states that were deterministically caused (or had their chances fixed) by prior mental and physical states outside of my control, plus the relevant laws. "I" become a stream of states/events in a causal chain that merely passes through me. Each member of the chain determines that the next member occurs.

In summary, then, property dualism denies libertarian freedom because it adopts either epiphenomenalism or event-event causation. Thus, mere property dualism, no less than physicalism, is false, given the truth of a libertarian account of free will, moral ability, moral responsibility, and punishment. Our commonsense notions about moral ability, responsibility, and punishment are almost self-evident. We all operate toward one another on the assumption that they are true (and these commonsense notions seem to assume libertarian free will). However, if physicalism or property dualism is true, we will have to abandon and revise our commonsense notions of moral ability, responsibility, and punishment because free will is ruled out.

SAMENESS OF THE SELF OVER TIME

Consider the following argument:

(1) If something is a physical object composed of parts, it does not survive over time as the same object if it comes to have different parts.
(2) My body and brain are physical objects composed of parts.
(3) Therefore, my body and brain do not survive over time as the same objects if they come to have different parts.
(4) My body and brain are constantly coming to have different parts.
(5) Therefore, my body and brain do not survive over time as the same objects.

(6) I do survive over time as the same object.
(7) Therefore, I am not my body or my brain.
(8) I am either a soul or a body or a brain.
(9) Therefore, I am a soul.

Premise (2) is commonsensically true. Premise (4) is obviously true as well. Our bodies and brains are constantly gaining new cells and losing old ones, or at least gaining new atoms and molecules and losing old ones. So understood, bodies and brains are in constant flux. I will assume that (8) represents the only live options for most people. This leaves the key premises (1) and (6).

Let's start with (1). Why should we believe that ordinary material objects composed of parts do not remain the same through part replacement?[11] To see why this makes sense, consider five scattered wooden boards, which we'll label a–e, each located in a different person's backyard. Commonsensically, it doesn't seem like the boards form an object. They are just isolated boards. Now, suppose we collected those boards and put them in a pile with the boards touching each other. We would now have, let us suppose, an object called a pile or heap of boards. The heap is a weak object indeed, and the only thing unifying it would be the spatial relationships between and among a–e. They are in close proximity and are touching each other. Now, suppose we took board b away and replaced it with a new board f to form a new heap consisting of a, c–f. Would our new heap be the same as the original heap? Clearly not, because the heap is just the boards and their relationships to each other, and we have a new board and a new set of relationships. What if we increased the number of boards in the heap to one thousand? If we now took one board away and replaced it with a new board, we would still get a new heap. The number of boards does not matter.

Now imagine that we nailed our original boards a-e together into a makeshift raft. In this situation, the boards are rigidly connected such that they do not move relative to each other; instead, they all move together if we pick up our raft. If we now took board b away and replaced it with board f, we would still get a new object. It may seem odd, but if we took board b away and later put it back, we would still have a new raft because the raft is a collection of parts and bonding relationships to each other. Thus, even though the new raft would still have the same parts (a–e), there would be new bonding relationships between b and the board or boards to which it is attached.

Now think of a cloud. From a distance, it looks like a solid, continuous object. But if you get close to it, say on a plane flight, it becomes evident that it is a very loose collection of water droplets. The cloud is like the heap of boards or the raft. If new droplets are added and some removed, it is, strictly speaking, not the same cloud.

Now, consider our bodies and brains, which are physical objects composed of billions of parts. From our daily vantage point, they appear to be solid, continuous objects. But if we could shrink down to the level of an atom, we would see that, in reality, they are like a cloud—gappy, largely containing empty space, and composed of billions of atoms (forming molecules and cells) that stand in various bonding relations between and among themselves. If we were to take a part away and replace it, we would have a new object. The body and brain are like the cloud or our raft. Besides the parts and the relationships among them, there is nothing in the body or brain to ground its ability to remain the same through part replacement. This is the fundamental insight behind the view that the body and brain cannot remain the same if there is part alteration.[12] Since the body and brain are constantly changing parts and relationships, they are not the same from one moment to the next in a strict philosophical sense (though, for practical day-to-day purposes, we regard them as the same in a loose, popular sense).

What about premise (6)? Why should we think we survive as the same object over time? Suppose you are approaching a brown table and in three different moments of introspection you attend to your own awarenesses or experiences of the table. At time t_1 you are five feet from the table and you experience a slight pain in your foot (P_1), a certain light brown table sensation (S_1) from a specific place in the room, and a specific thought that the table seems old (Th_1). A moment later at t_2 when you are three feet from the table you experience a feeling of warmth (F_1) from a heater, a different table sensation (S_2) with a different shape and slightly different shade of brown than that of S_1, and a new thought that the table reminds you of your childhood desk (Th_2). Finally, a few seconds later (t_3), you feel a desire to have the table (D_1), a new table sensation from one foot away (S_3), and a new thought that you could buy it for less than twenty-five dollars (Th_3).

In this series of experiences, you are aware of different things at different moments. However, at each moment of time, you are also aware that there is a self at that time that is having those experiences and that unites them into one field of consciousness. Moreover, you are also aware that the very same self had the experiences at t_1, t_2, and t_3. Finally, you are aware that the self that had all of the experiences is none other than you yourself. This can be pictured as follows:

ORIGINAL POSITION		**TABLE**
t_1	t_2	t_3
$\{P_1, S_1, Th_1\}$ →	$\{F_1, S_2, Th_2\}$ →	$\{D_1, S_3, Th_3\}$
I_1	I_2	I_3

$$I_1 = I_2 = I_3 = \text{I Myself}$$

Through introspection, you are aware that you are the self that owns and unifies your experiences at each moment of time and that you are the same self that endures through time. This is pretty obvious to most people. When one hums a tune, one is simply aware of being the enduring subject that continues to exist during the process. This is a basic datum of experience.

Moreover, fear of some painful event in the future or fear of blame and punishment for some deed done in the past appear to make sense only if we implicitly assume that it is literally I myself who will experience the pain or who was the doer of the past deed. If I do not remain the same through time, it is hard to make sense of these cases of fear and punishment. We would not have such fear or merit such punishment if the person in the future or past merely resembled my current self in having similar memories, psychological traits, or a body spatio-temporally continuous with mine or that had many of the same parts as my current body.[13]

Finally, some have argued that to realize the truth of any proposition or even entertain it as meaningful, the very same self must be aware of its different parts (e.g., those expressed by the associated sentence's subject, verb, and predicate). If one person-stage contemplated the subject, another stage the verb, and still another the predicate, literally no self would persist to think through and grasp the proposition as a whole.

For these and other reasons, we are warranted in believing that the I or self survives over time as the same object.[14]

If human beings have souls, then what are they like? How are spirit, mind, and will related to the soul? Are there such things as animal souls and, if so, are they different from the human soul? In the rest of this chapter we will tackle these questions.

THE HUMAN SOUL

The soul is a very complicated thing with an intricate structure. In order to understand that structure, we need to grasp two important issues: *the different types of states within the soul* and the notion of a *faculty of the soul*. The soul is a substantial, unified reality that informs (gives form to) its body. The soul is to the body like God is to space—it is fully "present" at each point within the body. Further, the soul and body relate to each other in a cause-effect way. For example, if I worry in my soul, my brain chemistry will change; if I will to raise my arm in my soul, the arm goes up. If I intake certain chemicals into my body (medication, etc.), my soul may be affected and my mood or feelings may change. The soul also contains various mental states within it, for example, sensations, thoughts, beliefs, desires, and acts of will. This is not as complicated as it sounds. Water can be in a cold or a hot state. Likewise, the soul can be in a feeling or thinking state.

FIVE STATES OF THE SOUL

Let us review the different states of consciousness that take place within the soul. As we have already seen, there are at least five different states contained in the soul. A *sensation* is a state of awareness, a mode of consciousness, such as a conscious awareness of sound or pain. A visual sensation like an experience of a tree is a state of the soul, not a state of the eyeballs. The eyes do not see. I (my soul) see with or by means of the eyes. A **thought* is a mental content that can be expressed in an entire sentence and that only exists while it is being thought. Some thoughts logically imply other thoughts. For example, "All whales are mammals" entails "This whale is a mammal." If the former is true, the latter must also be true. Some thoughts don't entail, but merely provide evidence for other thoughts. A **belief* is a person's view, accepted to varying degrees of strength, of how things really are. If a person has a belief (e.g., that it is raining), then that belief serves as the basis for the person's tendency to act on that belief (e.g., one gets an umbrella). At any given time, one can have many beliefs that are not currently being contemplated. A **desire* is a certain inclination to do, have, avoid, or experience certain things. Desires are either conscious or such that they can be made conscious through certain activities, for example, through therapy. An **act of will* is a volition or choice, an exercise of power, an endeavoring to do a certain thing, usually for the sake of some purpose or end.

FACULTIES OF THE SOUL

In addition to its states, at any given time the soul has a number of capacities that are not currently being actualized or utilized. To understand this, consider an acorn. The acorn has certain actual characteristics or states—a specific size or color. It also has a number of capacities or potentialities that could become actual if certain things happen. For example, the acorn has the capacity to grow a root system or change into the shape of a tree. Likewise, the soul has capacities. I have the ability to see color, think about math, or desire ice cream even when I am asleep and not in the actual states just mentioned.

Capacities come in hierarchies. There are first-order capacities, second-order capacities to have these first-order capacities, and so on, until ultimate capacities are reached. For example, if I can speak English but not Russian, then I have the first-order capacity for English as well as the second-order capacity to have this first-order capacity (which I have already developed). I also have the second-order capacity to have the capacity to speak Russian, but I lack the first-order capacity to do so. Higher order capacities are realized by the development of lower-order capacities under them. An acorn has the ultimate capacity to draw nourishment from the soil, but this can be actualized and unfolded only by developing the lower capacity to have a root system, then developing the still lower capacities *of* the root system.

The adult human soul has literally thousands of capacities

within its structure. But the soul is not just a collection of isolated, discrete, randomly related internal capacities. Rather, the various capacities within the soul fall into natural groupings called **faculties*. In order to get hold of this, think for a moment about this list of capacities: the ability to see red, see orange, hear a dog bark, hear a tune, think about math, think about God, desire lunch, desire a family. The ability to see red is more closely related to the ability to see orange than it is to the ability to think about math. We express this insight by saying that the abilities to see red or orange are parts of the same faculty—the faculty of sight. The ability to think about math is a capacity within the thinking faculty. In general, *a faculty is a "compartment" of the soul that contains a natural family of related capacities.*

We are now in a position to map out the soul in more detail. All of the soul's capacities to see are part of the faculty of sight. If my eyeballs are defective, then my soul's faculty of sight will be inoperative just as a driver cannot get to work in his car if the spark plugs are broken. Likewise, if my eyeballs work but my soul is inattentive—say, I am daydreaming—then I won't see what is before me either. The soul also contains faculties of smell, touch, taste, and hearing. Taken together, these five are called *sensory* faculties of the soul. The will is a faculty of the soul that contains my abilities to choose. The emotional faculty of the soul contains one's abilities to experience fear, love, and so forth.

MIND AND SPIRIT

Two additional faculties of the soul are of crucial importance. The **mind* is that faculty of the soul that contains thoughts and beliefs along with the relevant abilities to have them. It is with my mind that I think and my mind contains my beliefs. The **spirit* is that faculty of the soul through which the person relates to God (Ps. 51:10: Rom. 8:16; Eph. 4:23).[15] Before the new birth, the spirit is real and has certain abilities to be aware of God. But most of the capacities of the unregenerate spirit are dead and inoperative. At the new birth, God implants new capacities in the spirit. These fresh capacities need to be nourished and developed so they can grow.

ANIMAL SOULS

It is sometimes a surprise to people to learn that the Bible teaches that animals, no less than humans, have souls. In the Old Testament, *nephesh* (soul) and *ruach* (spirit) are used of animals in Genesis 1:30 and Ecclesiastes 3:21, respectively. In the New Testament, *psuche* (soul) is used of animals in Revelation 8:9. Moreover, it is a matter of common sense that animals are not merely unconscious machines. Rather, they are conscious living beings with sensations, emotions (like fear), desires, and, at least for some animals, thoughts and beliefs. The history of Christian teaching is widely united in affirming the existence of the "souls of men and beasts" as it has sometimes been put. But what is the animal soul like? Let us consider this question.

How do we decide what an animal's soul is like? Obviously, we cannot inspect it directly. We cannot get inside an animal's conscious life and just look at its internal states. The best approach seems to be this: Based on our direct awareness of our own inner lives, we should attribute to animals by analogy those states that are necessary to account for the animal's behavior, nothing more and nothing less.[16] For example, if a dog steps on a thorn and then howls and holds up its paw, we are justified in attributing to the dog the same sort of state that happens in us just after we experience such a stick. The dog feels pain. Now the dog may also be having thoughts about his unfortunate luck in stepping on the thorn, but there is no adequate evidence for this if we stick to what we observe about the dog's behavior. Such an attribution would be unjustified.

An interesting implication of this approach is that as we move down the animal chain to creatures that are increasingly unlike humans—from primates to earthworms—we are increasingly unjustified in ascribing a mental life to those animals. Now an organism either does or does not have a conscious life; for example, a worm either does or does not feel pain. But we have more grounds for ascribing painful sensations to primates than to worms according to the methodology above. All living animals have souls if they have organic life, regardless of the degree to which they are conscious, but we are justified in attributing less and less to the animal soul as the animal in question bears a weaker analogy to us.

In light of this methodology, what can we say about animal souls? Obviously, our answer will vary depending on the animal in question. But it seems reasonable to say that virtually all animals have certain sorts of sensations, for example, experiences of taste and pain. Many if not most animals seem to have desires as well, such as a desire for food. Many animals appear to engage in thinking and have certain sorts of beliefs. For example, a dog seems to be able to engage in means-to-ends reasoning. If he wants to go through a specific door to get food, and if the door is closed, he can select an alternative means to achieve the desired end. Many animals also engage in willings: that is, they will to do certain things, though there is no adequate evidence to suggest that they have libertarian freedom. It is more likely that an animal's will is determined by its beliefs, desires, sensations, and bodily states.

There are several capacities that animals do not seem to have. We have already mentioned libertarian freedom of the will. Animals also do not seem to have moral awareness. Animals do not seem to grasp key notions central to morality such as the notion of a virtue, of a duty, of another thing having intrinsic value and rights, of universalizing a moral judgment, and so on. They cannot distinguish between what they desire most and what is most desirable intrinsically. Alleged altruistic behavior can be explained on the basis of animal desire without attributing a sense of awareness of intrinsic duty to the animal.

Animals, therefore, do not seem to be capable of having a conflict between desire and duty, though they can experience a conflict between desires (e.g., to scratch the chair and to avoid being spanked). Animals do not seem to be able to entertain various sorts of abstract thoughts, for example, thoughts about matter in general or about love in general or even about food in general. Moreover, animals do not seem to be able to distinguish between true universal judgments (all alligators are dangerous) and mere statistical generalizations (most alligators are dangerous) nor do they have a concept of truth itself.

While this is controversial and I may be wrong in this

judgment, animals do not seem to possess language.[17] One problem that keeps people from getting clear about this is the presence of certain ambiguities about what language is. More specifically, the question of animal language cannot be adequately discussed without drawing a distinction between a sign and a symbol. A sign is a sense-perceptible object, usually a shaped thing like the characters "BANANA" or a sound (the utterance of "BANANA"). Now if an animal (or a human infant for that matter) comes to experience repeatedly the simultaneous presence of a sign (the visual presentation of BANANA) and the presence of a real banana, a habitual association will be set up such that the animal will anticipate the sense perception of a real banana shortly after seeing this shape: BANANA. In the case of the animal, BANANA does not represent or mean a banana, so it is not a symbol. Rather, BANANA is merely a certain geometrically perceived shape that comes to be associated with a banana in such a way that the latter is anticipated when the former is observed.

By contrast, real language requires symbols and not mere signs. When language users use the word *banana*, it is used to represent, mean, and refer to actual bananas. Now the evidence suggests that animals have certain abilities to manipulate and behaviorally respond to signs, but it is far from clear that they have a concept of symbols. One reason for this claim is the lack in animals of grammatical creativity and logical thought about language itself that is present in real language users.

Finally, St. Augustine once noted that animals have desires, but they do not have desires to have desires. They may have beliefs, volitions, thoughts, and sensations, but they do not seem to have beliefs about their beliefs, they do not choose to work on their choices, they don't think about their thinking, and they are not aware of their awarenesses. Nor do they seem to be aware of themselves as selves. In short, they do not seem to be able to transcend their own states and engage in reflection about their own selves and the states within them.

Animals are precious creatures of God and ought to be respected as such. But the animal soul is not as richly structured as the human soul, it does not bear the image of God, and it is far more dependent on the animal's body and its sense organs than is the human soul.

FINAL REFLECTION ABOUT THE RELEVANCE OF NEUROSCIENTIFIC DATA

In chapter 1, I claimed that the central issues involved in grasping the nature of consciousness and the soul are philosophical/theological and not scientific. Neuroscience is a wonderful tool for getting at the various causal interactions and dependency relations between brain and soul, but it is inept for resolving disputes about the nature and existence of consciousness and the soul. The central issues in those disputes include the philosophical, theological, and commonsense topics treated in chapters 2 through 4. Neuroscientific data are simply irrelevant for addressing those topics.

As a final illustration of the irrelevance of neuroscientific data for these topics, consider what we have learned about mirror neurons. These are neurons that respond during a motor activity and while an organism views another creature perform that same activity. For example, certain mirror neurons are activated in a chimpanzee's brain when it reaches for food, and those same neurons are activated when the chimpanzee watches another animal reach for food. Scientists think that mirror neurons are the neurological bases for various conscious activities such as experiencing empathy, comprehending the actions of others, anticipating action, and imitating others.

Now what exactly does the discovery of mirror neurons show regarding the central issues of the mind/body problem? Very little. One could interpret these causal, dependency data (e.g., our ability to empathize is causally dependent on healthy, functioning mirror-neuronal activity) in at least three empirically equivalent ways (ways that are consistent with the same scientific data such that those data cannot count in favor of one view over the others): (1) the feeling of empathy is identical to the mirror-neuronal activity (reductive physicalism); (2) the feeling of empathy is an irreducible mental property that emerges on mirror-neuronal activity (mere property dualism); (3) the feeling of empathy is an irreducible mental property that occurs in the soul and whose presence is causally dependent on proper mirror-neuronal activity in the brain (substance dualism).

My view is (3), but in any case, deciding among these three is not a scientific activity. After all, is it really so surprising to discover that various mental functions are causally dependent on a working body and brain? The ancients knew nothing about the brain, but they knew that if your eye is poked out, you will not be able to see, or if you get hit in the head, you get dizzy. But they didn't conclude from these facts that the eye sees or the brain has consciousness.

The causal dependency of feelings of empathy on mirror-neuronal activity is no different in kind from the dependence of sensations of sight on the eyes. It's just that today we have a more detailed story of the dependency, and that's really the only difference between the ancients and us.

In his 1886 lectures on the limitations of scientific materialism, John Tyndall claimed that "The chasm between the two classes of phenomena" (mental and physical phenomena) is of such a nature that we might establish empirical association between them, but it would still remain intellectually impassable. Let the consciousness of love, for example, be associated with a right-handed spiral motion of the molecules in the brain, and the consciousness of hate with a left-handed spiral motion. We should then know when we love that the motion is in one direction, and when we hate that the motion is in the other; but the "WHY" would remain as unanswerable as before.[18]

In my view, not much has changed since Tyndall penned these remarks. The details have changed, but the central issues have not.

CHAPTER IN REVIEW

- **A Case for Substance Dualism and the Immaterial Nature of the Self**
- Our Basic Awareness of the Self
- ☆ We are aware of our own self as being distinct from our bodies and from any particular mental experience we have, and as being an uncomposed, spatially unextended, simple center of consciousness. This grounds my properly basic belief that I am a simple

center of consciousness. In virtue of the law of identity, we then know that we are not identical to our body, but to our soul.

- Unity and the First-Person Perspective
  - ☆ If I were a physical object (a brain or body), then a third-person physical description would capture all the facts that are true of me. But a third-person physical description does not capture all the facts that are true of me. Therefore, I am not a physical object. Rather, I am a soul.
  - ☆ The unity of our conscious experience is best explained by an uncomposed mental substance, our soul.
- The Modal Argument
  - ☆ I am possibly disembodied (I could survive without my brain or body); my brain or body are not possibly disembodied (they could not survive without being physical); therefore, I am not my brain or body, I am a soul.
- Free Will, Morality, Responsibility, and Punishment
  - ☆ Property dualism and physicalism deny libertarian freedom. Property dualism attempts to explain human action either through epiphenomenalism or event-event causation. Yet, given the truth of a libertarian account of free will, moral ability, moral responsibility, and punishment, property dualism and physicalism are false. However, our commonsense notions about free will, moral ability, responsibility, and punishment, which are almost self-evident, are plausible given substance dualism.

- Sameness of the Self Over Time
  - ☆ A physical object composed of parts cannot survive over time as the same object if it comes to have different parts. My body and brain are physical objects composed of parts that are constantly changing, and therefore cannot survive over time as the same object. However, I do survive over time as the same object. Therefore, I am not my body or my brain, but a soul.

- **The Human Soul**
- Five States of the Soul
  - ☆ Sensation: A state of awareness, a mode of consciousness, e.g., a conscious awareness of sound or pain.
  - ☆ Thought: A mental content that can be expressed in an entire sentence and that only exists while it is being thought.
  - ☆ Belief: A person's view, accepted to varying degrees of strength, of how things really are.
  - ☆ Desire: A certain inclination to do, have, avoid, or experience certain things.
  - ☆ Act of will: A volition or choice, an exercise of power, an endeavoring to do a certain thing, usually for the sake of some purpose or end.
- Faculties of the Soul
  - ☆ Sensory faculties: Sight, smell, touch, taste, and hearing.
  - ☆ The will: A faculty of the soul that contains my abilities to choose.
  - ☆ Emotional faculties: One's abilities to experience fear, love, and so forth.

- Mind and Spirit
  - ☆ Mind: That faculty of the soul that contains thoughts and beliefs along with the relevant abilities to have them.
  - ☆ Spirit: That faculty of the soul through which the person relates to God.
- **Animal Souls**
- Animals have a soul, but it is not as richly structured as the human soul. It does not bear the image of God, and it is far more dependent on the animal's body and its sense organs than is the human soul.
- **Final Reflection about the Relevance of Neuroscientific Data**
- Neuroscience is a wonderful tool, but it is inept for resolving disputes about the nature and existence of consciousness and the soul. The central issues in those disputes include philosophical, theological, and commonsense topics. Neuroscientific data are simply irrelevant for addressing those topics.

KEY VOCABULARY

Act of will: A volition or choice, an exercise of power, an endeavoring to do a certain thing, usually for the sake of some purpose or end.

Belief: A person's view, accepted to varying degrees of strength, of how things really are.

Cartesian dualism: The mind is a substance with the ultimate capacities for consciousness, and it is connected to its body by way of a causal relation.

Desire: A certain inclination to do, have, avoid, or experience certain things.

Emergent dualism: a substantial, spatially extended, immaterial self emerges from the functioning of the brain and nervous system, but once it emerges, it exercises its own causal powers and continues to be sustained by God after death.

Epiphenomenalism: The mind is a by-product of the brain, which causes nothing; the mind merely "rides" on top of the events in the brain.

Event-event causation: The first event, combined with the laws of nature, are sufficient to determine or fix the chances for the occurrence of the second event.

First-person point of view: The vantage point that I use to describe the world from my own perspective.

Faculty of the soul: A "compartment" of the soul that contains a natural family of related capacities.

Mind: That faculty of the soul that contains thoughts and beliefs along with the relevant abilities to have them.

Properly basic belief: A belief that is not inferred from any other belief(s), but is rationally justified by experience (by perception, for example).

Spirit: That faculty of the soul through which the person relates to God.

Thomistic substance dualism: The (human) soul diffuses, informs (gives form to), unifies, animates, and makes human the body. The body is not a physical substance, but rather, an ensouled physical structure such that if it loses the soul, it is no longer a human body in a strict, philosophical sense.

Thought: A mental content that can be expressed in an entire sentence and that only exists while it is being thought.

NOTES

1. See Richard Swinburne, *The Evolution of the Soul*, rev. ed. (Oxford: Clarendon, 1997); *Mind, Brain & Free Will* (Oxford: Oxford University Press, 2013).
2. See J. P. Moreland and Scott Rae, *Body & Soul: Human Nature & the Crisis in Ethics* (Downers Grove, IL: IVP, 2000), chapter 6.
3. See William Hasker, *The Emergent Self* (Ithaca, NY: Cornell University Press, 1999).
4. Stewart Goetz, "Modal Dualism: A Critique," in *Soul, Body and Survival*, ed. Kevin Corcoran (Ithaca, NY: Cornell University Press, 2001), 89.
5. For more on the unity of consciousness, the binding problem, and split-brain phenomena, see Tim Bayne, "The Unity of Consciousness and the Split-Brain Syndrome," *The Journal of Philosophy* 105(6) (2008): 277–300; Tim Bayne and David Chalmers, "What Is the Unity of Consciousness?" in *The Unity of Consciousness*, ed. Axel Cleeremans (Oxford: Oxford University Press, 2003), 23–58. For an empirical argument against physicalism that centers on some of these considerations, see Eric LaRock, "An Empirical Case against Central State Materialism," *Philosophia Christi* 14(2) (2012): 409–26.
6. Hasker, *The Emergent Self*, 122–46.
7. Cf. Keith Yandell, "A Defense of Dualism," *Faith and Philosophy* 12 (1995): 548–66; Charles Taliaferro, "Animals, Brains, and Spirits," *Faith and Philosophy* 12 (1995): 567–81.
8. In normal life, I may be focusing on speaking kindly and be unaware that I am scowling. In extreme cases (multiple personalities and split brains), I may be fragmented in my functioning or incapable of consciously and simultaneously attending to all of my mental states, but the various personalities and mental states are still all mine.
9. For two reasons, quantum indeterminacy is irrelevant here: (1) The best interpretation of quantum indeterminacy may be epistemological and not ontological. (2) If quantum indeterminacy is real, events still have their chances fixed by antecedent conditions, and this is inconsistent with agent causation since on this view nothing fixes the chances of a free action.
10. Timothy O'Connor has argued that agent causal power could be an emergent property over a physical aggregate. See his *Persons and*

Causes: The Metaphysics of Free Will (New York: Oxford University Press, 2000). Subsequently, O'Connor has changed his view and opted for the idea that the agent is an emergent individual. See Timothy O'Connor and Jonathan D. Jacobs, "Emergent Individuals," *The Philosophical Quarterly* 53 (2003): 540–55. For a critique of O'Connor, see J. P. Moreland, *Consciousness and the Existence of God* (London: Routledge, 2008), chapter 4.

11. For more on problems of material composition, see Michael Rea, ed., *Material Constitution: A Reader* (Lanham, MD: Rowman & Littlefield, 1996); Christopher M. Brown, *Thomas Aquinas and the Ship of Theseus* (London: Continuum, 2005).
12. The view I am advancing is called *mereological essentialism* (from the Greek word *meros* that means "part"). Mereological essentialism is the idea that an object's parts are essential to its identity such that it could not sustain its identity to itself if it had alternative parts. Animalists and constitutionalists deny mereological essentialism. For a brief exposition of these views, see Eric Olson, *What Are We? A Study in Personal Ontology* (Oxford: Oxford University Press, 2007), chapters 2 and 3. In different ways, each view claims that, under certain circumstances, when parts come together to form a whole, as a primitive fact, the whole itself just is the sort of thing that can survive part alteration. In my view, this is just an assertion. The whole just is parts and various relations, and neither the parts nor the relations can sustain identity if alternatives are present. The whole is not a basic object—it is identical to its parts and relations.
13. Some claim that what unites all of one's various psychological stages into the life of one single individual is that the stages stand in an immanent causal relation to each other. But an immanent causal relation is one that holds between two states in the same thing. Thus, before a causal relation can be considered an immanent one, there must already be the same thing that has the two states. Because the immanent causal relation presupposes sameness of the thing in question, it cannot constitute what it is for the thing to be the same. Further, the immanent causal view confuses what it is that causes an object to endure over time with what it is for the object to remain the same.
14. For an additional argument for substance dualism, see Peter Unger, "The Mental Problems of the Many," in *Oxford Studies in Metaphysics I*, ed. Dean Zimmerman (Oxford: Clarendon Press, 2004), 195–222.
15. Biblical anthropological terms (heart, soul, spirit, mind) have a wide range of different meanings, and no specific use of a biblical term should be read into every occasion of the term. I am focusing here on a narrower, specific use of the term "spirit."
16. For more on this, see Richard Swinburne, *The Evolution of the Soul*, rev. ed. (Oxford: Clarendon Press, 1997), 11–16, 180–96, 200–219.

17. Cf. Swinburne, *The Evolution of the Soul*, 203–19; J. P. Moreland, ed., *The Creation Hypothesis* (Downers Grove, IL: IVP, 1994), chapter 7.
18. John Tyndall, "Scientific Materialism," in his *Fragments of Science Vol. II* (New York: P. F. Collier & Son, 1900), 95.

Chapter Five: THE FUTURE OF THE HUMAN PERSON

French philosopher Blaise Pascal once remarked that the immortality of the soul is something of such vital importance to us that one must have lost all feeling not to care about knowing the facts of the matter. Indeed. The Bible says that God placed eternity in our hearts. And nothing is more evident than that. Associated with the earliest archeological traces of the sons and daughters of Adam and Eve is the afterlife—depicted in cave drawings, rituals for burial, and so on. The pyramids stand as a testimony to our hunger for immortality. There is something in us that, unless it is repressed, reminds us regularly that this life is fleeting and that there must be more to it than our three score and ten.

If you think the afterlife is not real, it is not my purpose to provide evidence for it in this chapter. I have done that elsewhere.[1] I only note that if Jesus was raised from the dead, He has been to the afterlife and is qualified to tell us about it. And He assures us that there is a heaven and a hell. Further, if the biblical God exists, God values us who bear His image too much to let us pass out of existence. And He has a project/purpose He is working out on and for and through us. That project will never be finished, so He won't let us go.

The project—and we ourselves—means too much to Him. No competent artist destroys his work of art he likes while it is still in the process of being completed, and God is no different with respect to His unending work concerning us.

Finally, while the Bible is our ultimate source of authoritative information about heaven and hell, there is a growing, quite substantial body of evidence for heaven and hell from near-death experiences (NDEs) in which people become clinically dead, actually experience angels, demons, God, heaven or hell, and return to life. One must always be careful not to derive doctrine from such experiences, but the fact that they are real is, in my view, beyond question.

The Bible says that we die once and after that is the judgment (Heb. 9:27). But this truth does not rule out the reality of NDEs for two reasons. First, "death" in Hebrews 9:27 could mean "irreversible death," and, if so, it is not the same as clinical death. The latter allows for NDEs but not the former. Second, even if the passage just means "to die" without the notion of irreversibility, it still doesn't rule out NDEs. Why? Because throughout the Bible and in periods of church history, people such as Lazarus (John 11) died, continued to be in a conscious, intermediate state, and came back to life.

Here are two of many, many credible NDE accounts.[2] The first involves a woman named Kimberly Clark Sharp who worked at Harborview Hospital in Seattle. While she was attempting to resuscitate a clinically dead young patient—Maria—the patient suddenly became conscious, grabbed Kim's arm, and reported that she had left her body, floated out and above the hospital roof, and had seen an old large blue shoe with one part worn to the threads and with the laces tucked under the heel on an upper ledge of the hospital roof! The ledge was not accessible to anyone but hospital personnel, nor visible from buildings nearby, and Maria had never been to the hospital before. With her curiosity aroused by this bizarre story, Kimberly was shocked to find the shoe just as Maria had described in the location she had named.[3] Maria was interviewed by other witnesses that day who corroborate the incident.

A second, well-known account is about a woman named Viola who was checked into a hospital in Augusta, Georgia, in 1971 for routine gallbladder surgery. Six days after the surgery on May 5, her condition had worsened to the point that she was operated on again and died at 12:15 p.m. on the operating table. When the doctor said she was dead, Viola was confused. She had been in excruciating pain, when she suddenly felt a ring in her ear and, then, she popped out of her body! She found herself floating near the ceiling and gazed around the operating room, noting a number of things, including her own lifeless body. Though the room had been sealed off for surgery, she could hear voices in the outside hallway and passed through the wall where she saw her anxious family. Immediately, she noticed her daughter, Kathy, who was wearing an outfit Viola did not like. (Kathy had rushed to the hospital and put on the mismatched outfit hurriedly and without thinking.) She then noticed her brother-in-law talking to a family neighbor and saying, "Well, it looks like my sister-in-law is going to kick the bucket. I was planning on going to Athens, but I'll stick around now to be a pallbearer." Viola was infuriated by the insensitive comment.

She also sensed *presences* around her that she took to be

angels. And get this: She could travel anywhere her thoughts directed her, so she found herself instantaneously in Rockville, Maryland, where she saw her sister getting ready to go to the grocery store. Viola noted carefully the clothes her sister was wearing, her search for misplaced keys and a lost grocery list, and, finally, the car she drove. Moments later, she was whisked through a tunnel. Space forbids me to describe all she saw, but I must mention one thing. She met a baby who told Viola he was her brother. Viola was confused because she did not have a brother. The baby then showed himself to her dressed in quite specific clothing and told her that when she went back to tell her father about all this.

When Viola came back into her body, each and every detail I have shared was verified by the people involved, often with additional eyewitnesses. Viola's dad confirmed that only he, Viola's mother, and the doctor knew about the brother who had died as an infant but about whom the father and mother had decided to remain silent.

Make no mistake about it. The afterlife is real. In this closing chapter I want to do two things. First, I want to briefly say something about heaven and how important it is. Second, I will spend the bulk of the chapter describing the nature of hell and explaining why its existence is not only reasonable, but very important, and essential to God's purposes in the world.

HEAVEN IS A WONDERFUL PLACE

A human being is a functional unity of two distinct entities—body and soul. The human soul, while not by nature immortal (its immortality is sustained by God), is nevertheless capable of entering an intermediate disembodied state upon death, however incomplete and unnatural this state may be, and, eventually, being reunited with a resurrected body. Heaven will be a place where there is no longer any suffering, and life will be filled with joy, fulfillment, and a host of interesting, meaningful things to do.

I am not an expert at describing the nature of heaven, so I won't try. Happily, others have done a fine job of this.[4] Instead, I want to encourage you about how important heaven is. The cartoon character Charlie Brown once said, "I've developed a new philosophy of life. I only dread one day at a time."[5] Archibald Hart explains how many of us adopt this stance: "All of life is loss. . . . Life is all about loss. Necessary loss."[6] In point of fact, in this life we experience three kinds of losses. First, we often suffer the natural consequences of our own bad choices. We lose a job, a marriage, friends, health, and much more. And these losses stay with us. They are hard to shake.

Second, we suffer losses due to the fact that we live in a fallen, imperfect world. We grow old, lose our eyesight, our physical attractiveness (such as it may be!), athletic ability, loved ones due to death, our children to marriage, our sexual potency, and so forth. We discover that our marriage, career, and overall life satisfaction isn't what we hoped it would be. And as we age, we realize that many of our dreams will never come to pass.

Finally, there are several forms of injustice we all suffer that are never made right. From friends who gossip about us to bosses who bully us, to more severe crimes committed against us, the injustices of this life are not balanced and the harm done to us not completely healed.

This is where heaven becomes so very important. As Hart points out, a major problem with losses of various kinds is that we are *over*-attached—please note, I say over-attached—to this life and the things it offers: reputation, safety, our possessions, people who meet our deepest needs. Psychologists tell us that we need to have daily hope and optimism in life, and that such optimism must be rationally based so it isn't just a form of denial or a fantasy world out of touch with reality. In my view, to do this, we need to be able to place our losses—indeed, our entire, brief lives with all their attendant ups and downs—into the context of a broader, true, objectively meaningful picture. If we can do this, we can break our over-attachment to this life. The rational hope of heaven as the Bible presents it is just the sort of background belief one needs to navigate day-to-day life appropriately and with a proper perspective in assessing what it brings our way. And heaven gives us the rational hope that the injustices and other losses will, in fact, be made right.

This is no small deal. It's actually an essential perspective for living life well each day in God's kingdom. It's kind of ironic, really. People claim that the belief in heaven robs people of the value of life on this earth. But it's really just the opposite. In light of eternity, this life takes on incredible meaning and one has the perspective needed to live life well with the right priorities and the appropriate degree of attachments.

HELL: A TRAGIC YET AVOIDABLE OUTCOME

Hell is an unpleasant topic that we typically avoid. We may think about it from time to time, or even in our more fearful moments, talk about it with someone we trust. Some people even pull out the topic to scare someone into making a decision of faith. Others have brought it up and thrown it into a person's face, declaring damnation on that person for his harmful, disdainful acts.

But more often than not, the subject creeps up on us, silently, almost seditiously, until it enters our conscience and plagues us with thoughts of personal torment and judgment, or causes us to think about the agony a loved one will face if his life doesn't turn around. The topic can bring so much concern and pain into our lives that we put it into the back room of our minds and padlock the door. We do not want to talk about hell, much less dwell on it. We would rather the topic never came up.

But locking it away will not dispel its reality. In fact, throughout history, the concept of some kind of judgment and punishment for wrongdoing has played a critical role in religious life individually and communally. Certainly this has been true in the history of Judaism and Christianity—two of the most dominant religious forces in Western history.

In the Hebrew Scriptures, the prophet Daniel warned, "Multitudes who sleep in the dust of the earth will awake: some to everlasting life, others to shame and everlasting contempt" (Dan. 12:2 NIV). Turning to the New Testament, we find Jesus Himself advising His disciples not to "fear those who kill the body but are unable to kill the soul; but rather fear Him who is able to destroy both soul and body in hell" (Matt. 10:28). Then we see Paul, the great Christian missionary, proclaiming that "these will pay the penalty of eternal destruction, away from the presence of the Lord and from the glory of His power" (2 Thess. 1:9).

These leaders and teachers did not attempt to prove hell's reality. They knew it existed and warned their audiences appropriately. In prior periods of church history, there was a time when the reality of hell was so gripping, so pervasive, that the threat of excommunication was a powerful and feared danger.

Things have certainly changed. Today, hell is not a topic for polite conversations, and it rarely surfaces anywhere else, including sermons. We are afraid of it, embarrassed by it. Some even reject it as infantile and obnoxious. Atheist B. C. Johnson frankly states that "the idea of hell is morally absurd."[7] Morton Kelsey even notes that believers are ambivalent about the doctrine: "The idea of hell is certainly not popular among most modern Christians."[8]

THE SCRIPTURES ON HELL

Two New Testament passages provide the clearest definition of hell we have. Second Thessalonians 1:9 says, "These [who do not know God or obey the gospel] will pay the penalty of eternal destruction, away from the presence of the Lord and from the glory of His power." The other passage, Matthew 25:41 and 46, states: "Then He will also say to those on His left, 'Depart from Me, accursed ones, into the eternal fire which has been prepared for the devil and his angels'; . . . These will go away into eternal punishment, but the righteous into eternal life." From these (and other) verses we see that the essence of hell is the end of a road away from God, love, and anything of real value. It is banishment from the very presence of God and from the type of life we were made to live.

Hell is also a place of shame, sorrow, regret, and anguish. This intense pain is not actively produced by God; He is not a cosmic torturer. Undoubtedly, anguish and torment will exist in hell. And because we will have both body and soul in the resurrected state, the anguish experienced can be both mental and physical. But the pain suffered will be due to the shame and sorrow resulting from the punishment of final, ultimate, unending banishment from God, His kingdom, and the good life for which we were created in the first place.

Moreover, the flames in hell are most likely metaphorical. If metaphors for hell are taken literally, contradictions result. Hell is called a place of fire and darkness, but how could there be darkness if the fire is literal? Hell is described as a bottomless pit and a dump. How can it be both? In addition, Scripture calls God Himself a consuming fire (Heb. 12:29) and states that Christ and His angels will return surrounded "in flaming fire" (2 Thess. 1:8). But God is not a physical object as is fire, and the flames surrounding the returning Christ are no more literal than is the sword coming out of His mouth (Rev. 1:16). Flames are used as symbols for divine judgment.

SOME REASONS TO BELIEVE IN HELL

The severity and finality of the Bible's view of hell has been thought by some to be too horrendous, too absolute, too tragic to accept. It is no surprise, then, that the biblical picture has had many detractors. So now I want to examine the justification of hell by focusing on some of the issues, objections, and alternatives that have been proposed.

Oxford University philosopher Richard Swinburne has

offered an important defense of the orthodox view of hell.[9] Swinburne asks why it is that you have to have right beliefs and a good will (a properly directed will, one that desires God, salvation, and heaven) to go to heaven? Why are people with wrong beliefs and a bad will left out?

His basic answer is twofold: Heaven is the type of place where people with wrong beliefs and a bad will would not fit, and heaven must be freely and noncoercively chosen.

According to Swinburne, heaven is a place where people eternally enjoy a supremely worthwhile happiness. This happiness has three important aspects. First, it is not the mere possession of pleasant sensations. You could have pleasant sensations, say, by taking drugs all day or by having people constantly lie to you about how wonderful and intelligent you are. But for that you should be pitied. You would not have a supremely worthwhile happiness.

Second, such a happiness can only be possessed if you do what you truly want to without any conflicting desires. You could be happy doing something, even if you experienced conflicting desires about that activity, but it would be better to do something you freely wanted to do and it was free from conflict.

Third, a supremely worthwhile happiness must come from true beliefs and things that are truly and supremely valuable. We all know that happiness can be obtained from false beliefs. You can be happy in the belief that someone loves you even if that belief is false. So happiness can come from either true or false beliefs, but happiness is more worthwhile if it comes from true beliefs. If given a choice between a lot of happiness from false beliefs or a little happiness from true beliefs, we would choose the latter. Furthermore, happiness can come from doing silly or even immoral actions. Some people gain happiness from killing or stealing. But a supremely worthwhile happiness comes from true beliefs and activities that are really valuable.

To sum up then, a supremely worthwhile happiness is a deep happiness, not a shallow one. It does not involve the mere possession of pleasant sensations, but it is obtained by freely choosing to do activities when that choice is based on true beliefs and those activities are truly worthwhile. Deepest happiness is found in successfully pursuing a task of supreme value within a supremely valuable context when I have true beliefs about these and I only want to be doing these tasks in this situation without any conflict of my desires.

What are these supreme tasks and situations? Swinburne claims they include developing a friendship with God, learning to care for others who have that same friendship, caring for and beautifying God's creation, and the like. Heaven is not a reward for good behavior; it is a home for good people. Heaven intensifies and fulfills a certain type of life that can be chosen, in undeveloped form, in this life. Only people of a certain sort are suited for life in heaven: those who have a true belief about what it is like and really want to be there for the right reasons.

People with different beliefs about the good life or heaven will value and practice different activities, so even if they are seeking the good in some sense, their character will develop in a different way than will the character of the Christian. For example, a Buddhist who spends his whole life trying to remove his desires would not find heaven a place that fulfills the things he really wants. People with a bad will or people with a good will, but with false beliefs about what God, heaven, and the good really are, will not be suited for life in heaven.

Can God force the bad to become good? No, says Swinburne, if He respects our freedom. God can't make people's character for them, and people who do evil or cultivate false beliefs start a slide away from God that ultimately ends in hell. God respects human freedom, which He created. We could add here that it would be unloving, a form of divine coercion, to force people to accept heaven and God if they did not really want them. When God allows people to say no to Him, He actually respects and dignifies them. We may rush in to force our children to do something in their best interests, but our paternalism drops out when they grow up because we wish to respect them as adults. Similarly, God dignifies people and treats their choices as significant by allowing them to choose against Him, not just for Him.

So, Swinburne's argument is that heaven is suitable for people of a certain sort (those who really want to be there and who base their choice on true beliefs), and their decision to go there must be made freely. Hell is a place for people of a different character who freely choose to be there.

As it is, Swinburne's case seems to be a good one, but we can add to it too. For example, more can be said about how hell is the result of God's respect for persons. It is reasonable to argue that it is wrong to destroy the type of intrinsic value humans have. If God is the source and preserver of values, and if persons have the high degree of intrinsic value

Christianity claims they have, then God is the preserver of persons. He would be wrong to destroy something of such value just because it has chosen a life it was not intended to live. Thus, one way God can respect persons is to sustain them in existence and not annihilate them, as some claim will eventually happen to the lost. Annihilation destroys creatures of very high intrinsic value.

Another way to respect persons is to honor their free, autonomous choices, even if those choices are wrong. God respects persons in this second way by honoring their choices. As philosopher Eugene Fontinell has noted,

> The question that must be raised here is whether the doctrine of universal salvation, highly motivated though it may be, does not diminish the "seriousness" of human experience. . . . At stake here, of course, is the nature and scope of human freedom. . . . There is a profound difference between a human freedom whose exercise *must* lead to union with God and one that allows for the possibility of eternal separation from God. . . . A world in which there can only be winners is a less serious world than one in which the possibility of the deepest loss is real.[10]

Since God will not force His love on people and coerce them to choose Him, and since He cannot annihilate creatures with such high intrinsic value, the only option available is quarantine. And that is what hell is.

There are two other considerations to ponder concerning hell. First, some of God's attributes—particularly His justice and holiness—seem to demand the existence of hell.

Justice demands retribution, the distribution of rewards and punishments in a fair way. It would be unjust to allow evil to go unpunished and to reward evil with good. Thus hell is in keeping with God's justice. As Paul put it, "For after all it is only just for God to repay with affliction those who afflict you . . . dealing out retribution to those who do not know God and to those who do not obey the gospel of our Lord Jesus" (2 Thess. 1:6, 8).

Similarly, God's holiness requires Him to separate Himself entirely from evil, and hell is essentially a place away from God. Thus, hell is in keeping with God's holiness. It may very well be that our current hostilities toward the notion of hell result, not from an enlightened conscience, but quite the reverse. Our culture embraces a set of moral slogans devoid of content where individual liberties prevail, come what may. But as a society we have little concern or appreciation for holiness, and this dulling of our moral sensibilities may have inevitably led to our failure to appreciate the morality of hell as seen in the light of the demands of holiness and justice.

This matter leads to a second point. In ethics, there is a theory known as *virtue ethics*. The details of this theory are beyond my present concern, but one thing about virtue ethics is very important. This theory maintains that people who have a well-developed, virtuous character are in a better position to have moral sensibilities and genuine insight into what is right and wrong than those who do not have such a character. In other words, true moral experts are possible, and they are those people who have cared deeply about virtue and goodness and who have labored to develop ingrained virtues and moral sensibilities.

Jesus Christ and His apostles were moral experts. They were remarkable people who exhibited lives of staggering dedication to goodness, virtue, and the moral way of life. Now if they, being as virtuous as they were and having well-developed moral sensibilities, did not balk at the notion of hell but even embraced it as just, loving, and fair, then our current distaste for the doctrine says more about us than about the doctrine itself. To deny this conclusion is tantamount to claiming that our modern moral sensibilities are more developed than those of Jesus and His apostles, not to say those of the overwhelming number of godly people who have followed Jesus since. But this claim is clearly arrogant and unreasonable.

These, then, constitute some of the reasons the biblical doctrine of hell is morally and intellectually justifiable. But there are still some objections to consider.

SOME OBJECTIONS ANSWERED

The Problem of *Universalism

According to universalists, God will eventually reconcile all things to Himself, including all individuals, even if this means that God will continue to draw them to Himself in the afterlife. Morton Kelsey has said, "To say that men and women after death will be able to resist the love of God forever seems to suggest that the human soul is stronger than God."[11] John Hick claims that because of God's goodness, mercy, grace, and love, "God will eventually succeed in his purpose of winning all men to himself in faith and love."[12]

Universalists appeal to various arguments in support of their views, but three are central. One argument is that the doctrine of eternal banishment is immoral and unjust and, for that reason, unacceptable. Second, they argue that the doctrine of eternal banishment is incompatible with some of God's attributes, such as omnipotence, love, and mercy. God's mercy must surely triumph over human resistance. Finally, certain Bible texts (Acts 3:21; Rom. 5:18; 11:32; 1 Cor. 15:22–28; Eph. 1:10; 1 Tim. 2:4) are cited in favor of universalism. However, these arguments do not succeed. Let's look at each one more closely to see why.

We have already considered some reasons for rejecting the first argument—the injustice of eternal banishment. The state of hell is fair and, in fact, an indication of human dignity. Heaven is unsuited for certain types of people and lifestyles. People can freely resist God. God's love respects human freedom, making human choices and human history truly significant. And God will not extinguish people of intrinsic value. Eternal punishment is sad, even to God, but we must not confuse sadness with injustice. There are possibilities of real, eternal gains in this life, and this brings with it the possibility of real, eternal loss. And this latter possibility elevates the seriousness and significance of our world.

Regarding God's attributes, we can make a similar case. Omnipotence has nothing to do with the issue of hell. Consider the task of creating a square circle. This is a logical contradiction. The task of creating such an entity is a pseudo-task. It is not something you could do by, say, working out with weights. Power is irrelevant to such pseudo-tasks. The same can be said with regard to the free choices of human beings. All the power in the world cannot *guarantee* that a free choice will be a good one. *Determining* a good result of a *free* choice is a logical contradiction. Thus, although God is omnipotent, He still cannot do the logically impossible, including forcing humans (who possess divinely given free will) to make the right choices.

Furthermore, while hell is in some sense a defeat to God (His *desire* is that all men be saved), in another sense it is not a defeat. This is because hell is a quarantine that respects the freedom and dignity of God's image-bearers while separating the lost from His special presence and the community of those who love Him (heaven).

Finally, regarding divine love, we all know that resistance to love does not always break down. Love, even divine love, cannot coercively guarantee a proper response to it.

What about the Bible, then? Does it teach universalism? No. In fact, it contains very clear passages that contradict universalism (cf. Matt. 8:12; 25:31–46; John 5:29; Rom. 2:8–10; Rev. 20:10, 15). And the passages that appear to support universalism should be understood as doing one of two things. Either they are teaching what God's desire is without affirming that this will happen, or they are describing not the ultimate reconciliation of all of fallen humanity, but a restoration of divine order and rule over creation taken as a whole.

So universalism does not provide adequate grounds for rejecting belief in hell. As C. S. Lewis wisely observed,

> If a game is played, it must be possible to lose it. If the happiness of a creature lies in self-surrender, no one can make that surrender but himself (though many can help

> him to make it) and he may refuse. I would pay any price to be able to say truthfully, "All will be saved." But my reason retorts, "Without their will, or with it?" If I say, "Without their will," I at once perceive a contradiction: How can the supreme voluntary act of self-surrender be involuntary? If I say, "With their will," my reason replies, "How if they *will not* give in?"[13]

A SECOND CHANCE AFTER DEATH

The Bible is clear that people do not get a second chance to go to heaven after death. Hebrews 9:27 says, "It is appointed for men to die once and after this comes judgment." But is this teaching really fair and just? Yes. At least three factors tell us why.

For one thing, certain passages indicate that God gives people all the time they need to make a choice about eternity. Second Peter 3:9 teaches that God is postponing the return of Christ because He is "not wishing for any to perish but for all to come to repentance." From this, we can infer that if all a person needed were more time to make a decision, God would see to it that she got the extra time instead of dying prematurely. No one will go to hell who would have gone to heaven if he had just needed one more chance. Those who would have responded to a second chance after death will have their deaths postponed and be given that chance this side of the grave. God "desires all men to be saved and to come to the knowledge of the truth" (1 Tim. 2:4).

Second, people most likely would not have the ability to choose heaven after death, even if it were possible. Character is shaped moment-by-moment in the thousands of little choices we make. Each day our character is increasingly formed, and in each choice we make we either move toward or away from God. As our character grows, some choices become possible and others impossible. The longer one lives in opposition to God, the harder it is to choose to turn toward God. If God permits a person to die and go to hell, it seems reasonable to think that God no longer believes that this person is saveable. Only God could make that type of judgment, but that judgment could clearly be true.

And if God gives people a second chance after death, why did He create this world in the first place? Why not just go straight to a world in which everyone starts in the afterlife? Particularly, why did He not create people such that they start their lives in the conditions present in the afterlife if they were such that they would reject Christ in this life but (allegedly) would respond with a second chance after death? The second-chance idea makes this world superfluous, especially for those just described.

***Annihilationism**

Recently, some have argued for conditional immortality for the unsaved on Scriptural and moral grounds. Scripturally, it is claimed that biblical flames in hell are literal and that flames destroy whatever they burn. Morally, it is claimed that infinitely long punishing is disproportionate to a finite life of sin. Thus, everlasting punish*ment* in extinction is morally preferable to everlasting punish*ing*.

The scriptural argument is weak. Clear texts whose explicit intent is to teach the extent of the afterlife overtly compare the everlasting conscious life of the saved and the unsaved (Daniel 12:2; Matt. 25:41, 46). Moreover, as noted above, the flames in hell are most likely figures of speech for judgment and aren't intended to be taken literally (cf. Heb. 12:29; 2 Thess. 1:8).

The moral argument fails as well. For one thing, the severity of a crime is not a function of the time it takes to commit it. Thus, rejection of the mercy of an infinite God could quite appropriately warrant an unending, conscious separation from God. Further, everlasting hell is morally superior to annihilation as becomes evident from the following consideration.

Regarding the end of life and active euthanasia (the intentional killing of a patient), sanctity-of-life advocates eschew active euthanasia while quality-of-life advocates embrace it. The former reject it because on the sanctity-of-life view, one gets one's value, not from the quality of one's life, but the sheer fact that one exists in God's image. The latter accept it because the value of human life accrues from the quality of life. Thus, the sanctity-of-life position has a higher, not a lower, moral regard for the dignity of human life.

Now the traditional and annihilationist views about hell are expressions, respectively, of sanctity- and quality-of-life ethical standpoints. After all, the only grounds God would have for annihilating someone would be the low quality of life in hell. If a person will not surrender to God and if God will not extinguish one made in His image, then God's only alternative is quarantine, and that is what hell is. Thus, the traditional view, being a sanctity- and not a quality-of-life position, is morally superior to annihilationism.

What about Those Who Have Never Heard?

What about people who, through no fault of their own, never have a chance to hear the gospel of Christ? Do they receive or deserve unending punishing? Furthermore, why did God create people whom He knew would go to hell in the first place?

We must first affirm with Scripture that Jesus Christ is the only way to God. Christ is unique in His claims to be God (John 8:58; 10:30), to forgive sins (Mark 2:10), and in His miracles and resurrection from the dead. Buddha, Confucius, Mohammad, and other religious leaders are still in their graves; Jesus is not. Furthermore, Jesus Himself claimed to be the only way to God (John 3:18; 8:24; 11:25–26; 14:6) and this claim is reasserted by Peter in Acts 4:12.

The main issue in religion is truth, not belief. Believing something doesn't make it true. If four people have different beliefs about the color of my mother's hair, they can't all be right, and believing that her hair is red does not make it so. While all religions have certain truths in common, nevertheless, they significantly differ over what God is like, what God believes, what the afterlife is to be, and how we have a relationship with God. The real issue is truth, not belief. If Jesus was who He claimed to be, then He is unique and the only way to God.

We also need to observe that, according to the Bible, God desires all people to be saved (1 Tim. 2:4; 2 Peter 3:9; Ezek. 18:23, 32), and He judges fairly (Job 34:12; Gen. 18:25) and impartially (Rom. 2:11). The biblical God is not a cold, arbitrary being, but a God who deeply loves His creatures and desires their fellowship and worship.

Also, all humans have some light from creation and conscience that God exists, that He is personal and moral, and that they are guilty before Him (Rom. 1:18–20; 2:11–16).

Moreover, the Bible is very clear about the state of those who hear the gospel and reject it (John 3:18; 5:21–24). They will be barred from heaven and sent to judgment in hell. Remember, the most kind, virtuous person who ever lived said these words.

With all this in mind, we can begin to address the first question raised: What about those who don't have a chance to hear the gospel? The Bible doesn't address this question explicitly and for obvious reasons. The Word of God doesn't usually offer a plan B if the church chooses to reject God's plan A. Scripture commands us to go into the world and be sure no one fails to hear the gospel. It doesn't explicitly say, "Here is what will happen if you decide not to act on God's command." So whatever view we reach here must be formulated theologically from God's attributes and general considerations in Scripture.

Here is another point: We must distinguish between the *means* of salvation and the *basis* of salvation. Christ's death and resurrection have always been the basis for our justification before God. However, the *means* of appropriating that basis has not always been a conscious knowledge of the content of the gospel. Saved individuals before Christ (and surely justice includes people who lived and died within a few years after Christ's execution when the gospel couldn't reach them) were saved on the basis of Christ's work, but they did not know the content of the gospel. They were saved by responding in faith to the revelation they had received at that point (Gen. 15:6).

Furthermore, most theologians believe that those who cannot believe (infants and those without rational faculties capable of grasping the gospel) have the benefits of Christ applied to them. Many argue this on the basis of 2 Samuel 12:23, where David expresses his conviction that he will be reunited with his deceased infant in the afterlife. They also appeal to the fact that there is no mention of perdition for children in all the Bible, and they cite God's clear desire to save all humanity, His justice, and His love.[14]

So I believe it is certainly possible that those who are responding to the light from nature that they have received will either have the message of the gospel sent to them (cf. Acts 10) or else it may be that God will judge them based on His knowledge of what they would have done had they had a chance to hear the gospel. The simple fact is that God rewards those who seek Him (Heb. 11:6). It does not seem just for another to be judged because of my disobedience in taking the gospel to others, and it is surely the case that the gospel has not been taken to others in the way God commanded. I am not sure this line of reasoning is true, but some deem it plausible in light of the information we have. However, at the end of the chapter I will offer another view of the fate of the unevangelized that is at least as plausible if not more so than the one we are now considering.

If this case is correct, then why reach out to others with the gospel? For three reasons: As I admitted, this answer is somewhat speculative (remember, the Bible does not address the question explicitly). While I think it could be right, we should evangelize just in case it is wrong. Also, God commands us to tell others about His Son, and we should obey Him out of our love for Him. We are also told to spread God's teachings and broaden His family, not merely for what happens in the future state, but to spread His rule now. Why delay and give evil more victory in the present? Why not bring people mercy and forgiveness and release from sin as soon as possible? Good news should not be delayed.

New Testament scholar Leon Morris puts this whole discussion in perspective:

> Peter told [Cornelius] that God is no respecter of persons, "But in every nation he who fears him and works righteousness is acceptable to him" (Acts 10:35). This surely means that people are judged by the light they have, not by the light they do not have. We remember, too, that Paul says, "It is accepted of a man according to what he has and not according to what he does not" (2 Cor. 8:12). Long ago Abraham asked, "Shall not the Judge of all the earth do right?" (Gen. 18:25), and we must leave it there. We do not know what the fate of those who have not heard the gospel will be. But we do know God, and we know that he will do what is right.[15]

But a final question remains: Why did God create people whom He knew would not choose Him? In my view, Christian philosopher William Lane Craig has provided a very helpful answer to this question.[16]

According to Craig, among the things God knows is His knowledge of what every possible free creature would do under any possible set of circumstances. This is sometimes called *"middle knowledge." This is knowledge of those creatable worlds God can actually create. For example, I do not have a sister, but God knows a possible person who would have been my sister if my parents had married earlier and given birth to a daughter. Again, I was raised in Missouri and never decided to become a lawyer, but God knows what would have happened if I had moved to Illinois as a teenager and what I would have freely done had I been challenged as a boy to become a lawyer. This is knowledge of what a free creature would do in certain circumstances, even if those circumstances do not happen. God knows all the possible creatures He could have created but didn't, and He knows all the free choices all His creatures—those He actually created and those He did not create—would make in all the circumstances they could be placed in (some actually happening, some not happening).

What does this have to do with the doctrine of hell? God knows every possible creature and every possible response they would make to the gospel in every possible circumstance. Given this knowledge, why did God create a world in which people are not saved (He knew before they were born that they would not trust Christ)? Furthermore, because God knows what circumstances need to happen for each person to trust Christ, why didn't God bring those circumstances about instead of other circumstances such that persons placed in them freely reject Christ?

Craig breaks this problem down into four statements:

(1) God has middle knowledge.

(2) God is omnipotent (all-powerful).
(3) God is all-loving.
(4) Some persons freely reject Christ and are lost.

According to Craig, the problem is this: If we accept the first three statements, an objector would claim that we cannot also accept statement 4. For if we accept statements 1–3, the objector holds that we also ought to accept these statements:

(1') God knows under what circumstances any possible person would freely receive Christ.
(2') God is able to create a world in which all persons freely receive Christ.
(3') God holds that a world in which nobody rejects Christ is preferable to a world in which somebody does and consequently is lost.

The objector claims, then, that since God has middle knowledge, He would know for every possible creature just what circumstances need to happen to bring him to Christ, and since God prefers a world in which nobody rejects Christ over a world in which some reject Christ, then God would have the knowledge and power to create a world in which everyone is saved.

Craig's solution to this problem is to reject 1'–3' and replace them with these statements that are more likely to be true:

(1") There are some possible persons who would not freely receive Christ under any circumstances.

(2”) There is no possible world in which all persons would freely receive Christ.

(3”) God holds that a world in which some persons freely reject Christ but the number of those who freely receive Him is maximized is preferable to a world in which a few people receive Christ and none are lost.

Let us look at these in more detail. We have already discussed 1” in conjunction with universalism. There we saw that God cannot guarantee that a free creature would accept Christ. That is just what it means to be free. Therefore, of all the possible persons God could have created or did create, some would freely reject Christ no matter what the circumstances. How could God *guarantee* a set of circumstances for each person in which that person *freely* receives Christ? Statement 1” seems clearly true then.

For all we know, of all the possible persons God could have created, the vast majority of those who would have rejected Christ never get created in the first place. The number of people who reject Christ may be an act of mercy on God’s part. But still, Craig reminds us, the objector may respond by asking why God created *anyone* whom He knew would not trust Christ.

Craig’s answer is 2”. Perhaps there is no world God could have created in which all persons freely receive Christ. Now, on the surface of it, 2” does not seem plausible. Suppose of all the possible persons God could have created (including some He did create and some He did not create), there is a set *n* composed of all and only those people who would trust

Christ. Then why couldn't God just create a world composed only of people in set *n?* What is the problem here?

Craig replies as follows: It may not be possible to create just those persons and just the right circumstances for all to be saved. Why? It may well be that if God changes the circumstances that allow Smith to freely trust Christ, this alteration may bring it about that Jones will freely reject Christ even though Jones would have accepted Christ in a world without the circumstances needed to bring Smith to saving faith.

An example may help to illustrate this point. Suppose God can bring about two circumstances, one in which my father is offered a job in Illinois while I am a young boy, and one in which no such offer is made. In the former case, suppose my father freely accepts the offer and we move to Illinois. In the latter case, we stay in Missouri. Let us call these events *C* and *D*, respectively. Suppose further that in circumstance *D*, three years after the offer could have been given (but wasn't), I will meet just the right person in just the right circumstances and come to Christ. It is entirely possible that I would have had no such opportunity in circumstance *C*. So my salvation is dependent upon *D* obtaining as opposed to *C*. In addition, suppose that if *D* obtained, I would lead five others to Christ in Missouri in my lifetime, but if *C* had obtained, then a neighbor of mine in Illinois would have come to Christ by watching my non-Christian life fall apart, but without my bad example he would freely reject Christ. Now suppose this neighbor would have eventually led ten people to Christ. In circumstance *D*, six people come to Christ (I and five others), and in *C*, eleven come to Christ. *C* and *D* cannot both obtain, and thus free human choices responding to different influences make it impossible for God to bring about the conversion of all seventeen people.

This example shows that adjusting the circumstances in a possible world has a ripple effect. Not even God can change things piecemeal and respect freedom. If one thing is changed, this has an impact on other things. Additionally, the more people God creates, the greater the chance that some of the people He makes will not trust Christ. So 2" seems reasonable and quite plausible.

There is another point that can strengthen 2". In the ancient church there were two major views about the origin of the soul: *creationism and *traducianism. According to creationism, our bodies are passed on to us through normal reproduction by our parents, but God creates each individual soul out of nothing, most likely at fertilization. According to traducianism, both the body and soul are passed on to us by our parents. Now the soul is the thing that makes us the unique individuals we are. I could have had a different body, but I could not have had a different soul. My soul makes me, *me*.

For the creationist, I could have had different parents from the ones I had. Why? Because God could have created my soul out of nothing and placed it into a different body formed by different parents. In this case, I would have been united to a different body and born to different parents. For the traducian, I could not have had different parents from the ones I had. Why? Because essential to my identity is the fact that I have this very soul, and essential to a particular soul's being the very soul it is, is that it come from just these two people. The soul is passed on from the parents—different parents, different soul.

If we accept traducianism, then God could not have created me without creating my specific parents, and He could not have created my specific parents without creating their specific parents, and so on. In other words, God could only get to me, as it were, by reaching me through my entire ancestral chain. If my great-grandparents had married different people, I could not have existed. So when God is comparing alternative possible worlds, He is not just comparing alternative individuals, but alternative ancestral chains in their entirety. It may be that God allows some chains to come about, with some individuals in them who reject Christ (say my great, great-grandfather), but which allow for others to be born who do trust Christ. In this case, God would be balancing alternative chains and not just alternative people. Of course, if one accepts creationism regarding the soul (not to be confused with creationism as opposed to evolution), then this solution would be unavailable.

These considerations show that creating a world with a large number of people may have the result that a number of them may be permitted to be lost in order to respect human freedom and accomplish some task known by God. What might that task be? Statement 3" gives us an answer: God prefers a world in which some persons freely reject Christ but the number of saved is maximized, over a world in which a few trust Christ and none are lost.

Consider two worlds, W1 and W2. In W1, suppose fifty million are saved and five million are lost, while in W2, five million are saved and none are lost. It is not clear that W2 is morally preferable to WI. If W2 *is* morally preferable, then hell has veto power over heaven. God's purpose becomes the negative one of keeping people from hell, not the positive one of getting people to heaven. In contrast, it may be worth having more people go to heaven to allow more to go to hell. At the least, this is not clearly immoral. If something like this is correct, then, with Craig, we can affirm the following and add it to his opening four statements:

> (5) The actual world contains an optimal balance between saved and unsaved, and those who are unsaved would never have received Christ under any circumstances.

This would seem to explain why God would create individuals whom He knew would not trust Christ in any circumstances.

Our discussion of middle knowledge has been conducted with a view toward solving the problem of why God created people whom He knew would not choose Him. Before we close the chapter, we want to relate middle knowledge to a closely connected question discussed a few pages earlier about the fate of people who never have a chance to hear the gospel message. In our earlier discussion we suggested a possible answer to this question: either God will get the gospel message to such people or else He will judge them based on His knowledge of what they would do if they had been given a chance to hear the gospel. We now offer a third alternative solution to the question that is at least as likely to be true as the other options.

According to the third solution, no one who does not hear the gospel message and accept it will be saved. Moreover, this solution can be clarified by taking it to imply the following two (apparently conflicting) theses:

(6) In unevangelized areas, there will be people who will be saved if someone were to take the gospel to them who would not be saved if no one took the gospel to them.
(7) In unevangelized areas, there will be no one who will go to hell who would have accepted the gospel if someone would have taken it to them.

Proposition (6) assures us that if we take the gospel to unevangelized areas, there will be people there who will be saved who would otherwise not have been saved and (7) expresses the idea that no one who does not get a chance to hear the gospel and is lost would have trusted Christ if they had been given the chance. But how can both (6) and (7) be true? How can it be the case that if we go to an unreached people, there will be people saved who otherwise would not be saved while at the same time accepting the idea that if no one takes the gospel to that unreached people group, no one will be lost who would have been saved had someone gone to them?

The answer involves God's middle knowledge. Suppose that God is contemplating some people group through its history and He is deciding whom He will and will not create among all the possible persons He could create there. Surely, the number of people in the history of the people group under consideration could have been larger or smaller than the actual number that obtains in the real world. There are possible persons God could have created and placed in this people group but for some reason or another are not brought into the world. Now suppose that among the possible persons God knows He could create in this people group, there is some specific set of possible persons who would respond to the gospel if they were given the chance. God would know the number of those possible persons and He would know who they were. Let us use the name "set A" to refer to the set of possible persons in this people group who would trust Christ if given the chance.

Now God is deciding who to create in the people group under consideration. Should He create the people in set A or not? It depends on whether or not someone is willing to take the gospel to the people group in question. If God knows someone will bring the gospel, He will create the people in set A and there will be people saved (those in set A) who would not be saved if no one was willing to go to the people group; yet, if no one is willing to bring the gospel, God will refrain from creating the people in set A, knowing that they would never be saved. This is what (6) means. If no one is willing to bring the gospel to this people group, then the only people God will allow to be created are unsaveable people, those who would not trust Christ even if they had a chance to hear about Him. This is what (7) means. God's choice to create the people in set A depends on whether or not someone will freely go and preach the gospel to the relevant people group. This, then, is a third solution to the question of the unevangelized that we are considering. In sum, the unevangelized on this view are not lost because they never have a chance to hear and respond to the gospel, but because God knew in advance that they would not respond (even if given the chance), and thus placed them in their specific geographical location. As a result, their salvation does not depend on contingencies of history over which they have no control (missionaries being sent), and God is just in His actions toward them. Instead, God gives anyone He knows would respond the opportunity to hear the gospel, and thus places them in circumstances where they can in fact hear and respond. Consequently, no one is lost whom God knows would accept the gospel if given the opportunity.

In this closing chapter, I have tried to share with you some important facts about the reality and nature of heaven and hell. I can't think of a better summary of these things than the one offered by the apostle Paul (which I paraphrase): "For the wages of sin is death—eternal separation from God—but the free gift of God is an eternal kind of life that begins now and lasts forever in Christ Jesus our Lord" (Rom. 6:23).

CHAPTER IN REVIEW

In this final chapter we looked at what the future holds for the human soul in the afterlife. Scripture has much to say about heaven and hell, and near-death experiences provide extrabiblical confirmation that the soul survives the death of the body. We saw that heaven provides the Christian with a great source of rational hope for the future, and that hell is a tragic but avoidable final destiny. Although many have challenged the morality of hell, it makes sense both in relation to God's attributes and to human character, which can develop in ways that are antithetical to heaven. We also discussed a number of alternatives to the traditional view of hell, and ended by considering the destiny of the lost and the unevangelized and how we should understand this in relation to God's justice.

- **The Future of the Human Person**
- Because Jesus was raised from the dead, and has been to the afterlife, He is qualified to tell us about it. And He assures us that there is a heaven and a hell.
- God values us who bear His image too much to let us pass out of existence.
- Near-death experiences provide significant evidence for heaven and hell.

- **Heaven Is a Wonderful Place**
- Heaven will be a place where there is no longer any suffering, and life will be filled with joy, fulfillment, and a host of interesting, meaningful things to do.
- Heaven offers us rational hope and optimism in life. We place our losses—indeed, our entire, brief lives with all their attendant ups and downs—into the context of a broader, true, objectively meaningful picture.

- **Hell: A Tragic Yet Avoidable Outcome**
- 2 Thessalonians 1:9 and Matthew 25:41, 46: the essence of hell is banishment from the very presence of God and from the type of life we were made to live.
- Some Reasons to Believe in Hell
- ☆ Swinburne's first argument: Heaven is the type of

place where people with wrong beliefs and a bad will would not fit.

☆ Swinburne's second argument: Heaven must be freely and noncoercively chosen, which makes hell a real possibility for some.

☆ Hell is the result of God's respect for the intrinsic value of humans as well as their free, autonomous choices.

☆ God's justice demands a fair distribution of rewards and punishments. It would be unjust to allow evil to go unpunished and to reward evil with good.

☆ God's holiness requires Him to separate Himself entirely from evil, and hell is essentially a place away from God.

☆ Jesus and His apostles were remarkable people dedicated to goodness and virtue. If they did not balk at the notion of hell but embraced it as just, loving, and fair, then our distaste for the doctrine says more about us than about the doctrine itself.

- **Some Objections Answered**
- The Problem of Universalism

☆ The state of hell is fair and, in fact, an indication of human dignity.

☆ Love cannot coerce.

☆ Scripture contradicts universalism (Matt. 8:12; 25:31–46; John 5:29; Rom. 2:8–10; Rev. 20:10, 15).

- A Second Chance after Death

☆ Scripture indicates that God gives people all the time

they need to make a choice about eternity (2 Pet. 3:9; 1 Tim. 2:4).

☆ If God permits a person to die and go to hell, it seems reasonable to think that God no longer believes that this person is saveable, perhaps because of their formed character.

- Annihilationism

☆ Clear texts whose explicit intent is to teach the extent of the afterlife overtly compare the everlasting conscious life of the saved and the unsaved (Dan. 12:2; Matt. 25:41, 46).

☆ The severity of a crime is not a function of the time it takes to commit it. Thus, rejection of the mercy of an infinite God could quite appropriately warrant an unending, conscious separation from God.

☆ If a person will not surrender to God and if God will not extinguish one made in His image, then God's only alternative is quarantine and that is what hell is. Thus, the traditional view, being a sanctity- and not a quality-of-life position, is morally superior to annihilationism.

- What about Those Who Have Never Heard?

☆ The Bible is not explicit on this issue, but we can be confident that God is just in all that He does. One approach to this question is to understand that the world contains an optimal balance between saved and unsaved, and those who are unsaved would never have received Christ under any circumstances.

KEY VOCABULARY

Annihilationism: The view that immortality is conditional and that some will be completely destroyed, as opposed to existing in eternal punishment in hell.

Creationism: Our bodies are passed on to us through normal reproduction by our parents, but God creates each individual soul out of nothing, most likely at fertilization.

Middle knowledge: God's knowledge of what creatures would freely do in any given set of circumstances.

Traducianism: Both the body and soul are passed on to us by our parents.

Universalism: The view that God will eventually reconcile all things to Himself, including all individuals, even if this means that God will continue to draw them to Himself in the afterlife.

NOTES

1. Gary Habermas and J. P. Moreland, *Beyond Death* (Wheaton, IL: Crossway Books, 1998; rev. ed. of *Immortality: The Other Side of Death* [Nashville: Thomas Nelson], 1992).
2. Both are from Peter Shockey, *Reflections of Heaven* (New York: Doubleday, 1999), 147–48 and 163–72, respectively. For an interesting story of an atheistic university professor who died, went to hell, came back, left teaching and is now in the ministry, see Howard Storm, *My Descent Into Hell* (New York: Doubleday, 2005). There are a number of details reported by Storm with which I disagree theologically, but the general account itself seems to me to be credible.
3. See Habermas and Moreland, *Beyond Death*, 212–14, for more on Maria's case.
4. For example, Randy Alcorn, *Heaven* (Carol Stream, IL: Tyndale, 2004).
5. As quoted in Archibald D. Hart, *Unmasking Male Depression* (Nashville: Thomas Nelson, 2001), 125.
6. Ibid.

7. B. C. Johnson, *The Atheist Debaters Handbook* (Buffalo, NY: Prometheus, 1981), 116.
8. Morton Kelsey, *Afterlife: The Other Side of Dying* (New York: Paulist, 1979), 237.
9. Richard Swinburne, *Faith and Reason* (Oxford: Clarendon, 1981), 143–72; *Responsibility and Atonement* (Oxford: Clarendon, 1989), 179–200; "A Theodicy of Heaven and Hell," in *The Existence & Nature of God*, ed. Alfred J. Freddoso (Notre Dame, IN: University of Notre Dame, 1983), 37–54.
10. Eugene Fontinell, *Self, God, and Immortality* (Philadelphia: Temple University, 1986), 217. Cf. a review of this work by J. P. Moreland, *International Philosophical Quarterly* 29 (1989), 480–83.
11. Morton Kelsey, *Afterlife*, 251.
12. John Hick, *Evil and the God of Love* (New York: Harper & Row, 1978), 342; cf. *Death and Eternal Life* (San Francisco: Harper & Row, 1980), 242–61.
13. C. S. Lewis, *The Problem of Pain* (New York: Macmillan, 1962), 106ff.
14. See Robert Lightner, *Heaven for Those Who Can't Believe* (Schaumburg, IL: Regular Baptist, 1977).
15. Leon Morris, "The Dreadful Harvest," *Christianity Today*, May 27, 1991, 34–38.
16. See William Lane Craig, "'No Other Name': A Middle Knowledge Perspective on the Exclusivity of Salvation through Christ," *Faith and Philosophy* 6 (1989), 172–88; *The Only Wise God* (Grand Rapids, MI: Baker, 1987), 127–52; *No Easy Answers* (Chicago: Moody, 1990), 105–16.

GLOSSARY

This glossary defines terms found throughout the text that are preceded by an asterisk (*) symbol.

Abstract: In the discipline of philosophy, this term refers to properties (e.g., redness, hardness; see also the entry for "property" below) and relations (e.g., taller than, heavier than) that do not exist by themselves in space or time, but can exist potentially *in* many different places and times (e.g., *redness* can exist *in* both an apple and a ball; a dog can be *heavier than* a rock, while a rock can be *heavier than* a leaf). This term is often contrasted with spatial/temporal objects that are *concrete* or *physical* (e.g., a house, a cow, a bracelet).

Act of will: A volition or choice, an exercise of power, an endeavoring to do a certain thing, usually for the sake of some purpose or end.

Annihilationism: The view that immortality is conditional and that some will be completely destroyed, as opposed to existing in eternal punishment in hell.

Anti-Cartesian principle: There can be no purely mental beings because nothing can have a mental property without having a physical property as well.

Belief: A person's view, accepted to varying degrees of strength, of how things really are.

Cartesian dualism: The mind is a substance with the ultimate capacities for consciousness, and it is connected to its body by way of a causal relation.

Causal reduction: The causal activity of the reduced entity is entirely explained in terms of the causal activity of the reducing entity.

Consciousness: Broadly, what you are aware of when you engage in first-person introspection.

Creationism: Our bodies are passed on to us through normal reproduction by our parents, but God creates each individual soul out of nothing, most likely at fertilization.

Desire: A certain inclination to do, have, avoid, or experience certain things.

Dualism: The view that the soul is an immaterial thing different from the body and brain.

Eliminative materialism: Mental terms get their meaning from their role in folk psychology, and will eventually be replaced with some neurophysiological theory.

Emergent dualism: a substantial, spatially extended, immaterial self emerges from the functioning of the brain and nervous system, but once it emerges, it exercises its own causal powers and continues to be sustained by God after death.

Emergent supervenience: The view that mental properties are distinctively new kinds of properties that in no way characterize the subvenient physical base on which they depend.

Epiphenomenalism: The mind is a by-product of the brain, which causes nothing; the mind merely "rides" on top of the events in the brain.

Epiphenomenon: Something that is caused to exist by something else but that itself has no ability to cause anything.

Event: a temporal state that occurs in the world (e.g., water freezing or a dog barking).

Event-event causation: The first event, combined with the laws of nature, are sufficient to determine or fix the chances for the occurrence of the second event.

Extinction/re-creation view: When the body dies the person ceases to exist since the person is in some sense the same as his or her body. At the future, final resurrection, persons are re-created after a period of non-existence.

Faculty of the soul: A "compartment" of the soul that contains a natural family of related capacities.

First-person point of view: The vantage point that I use to describe the world from my own perspective.

Folk psychology: A common sense theory designed to explain the behaviors of others by attributing mental states to them.

Functionalism: The physicalist view that reduces mental properties/states to bodily inputs, behavioral outputs, and other mental state outputs.

Immediate resurrection view: At death, in some way or another, each individual continues to exist in a physical way.

Individual ontological reduction: One object (a macro-object like a dog, a molecule, or a person) is identified with another object or taken to be entirely composed of parts characterized by the reducing sort of entity.

Intentionality: The "of-ness" or "about-ness" of various mental states.

Knowledge: To represent reality in thought or experience the way it really is on the basis of adequate grounds.

Knowledge by acquaintance: Knowledge of a thing when one is directly aware of that thing.

Know-how: The ability to do certain things.

Linguistic reduction: One word or concept (*pain*) is defined as or analyzed in terms of another word or concept (*the tendency to grimace when stuck with a pin*). These kinds of reductions are definitional.

Mental holism: The notion that a given mental state gets its identity from its entire set of relations to all the other mental states in one's total psychology.

Metaphysics: In philosophy, this refers to the study of the most fundamental aspects of reality that underlie what we experience through our senses. Common topics of study in metaphysics include existence, substance, properties, causation, events, and mind/body questions.

Middle knowledge: God's knowledge of what creatures would freely do in any given set of circumstances.

Mind: That faculty of the soul that contains thoughts and beliefs along with the relevant abilities to have them.

Mind-body dependence: What mental properties an entity has depend on and are determined by its physical properties.

Mind-body problem: The problem of understanding the relationship between the apparently immaterial mind and the physical body and brain.

Ontology (Ontological): A branch of metaphysics that deals with the nature of being and existence. Ontological questions include whether humans possess a soul, and whether abstract entities such as numbers truly exist.

Physicalism (or strict physicalism): The view that the only things that exist are physical substances, properties, and events. In relation to humans, the physical substance is the material body, especially the brain and central nervous system.

Properly basic belief: A belief that is not inferred from any other belief(s), but is rationally justified by experience (by perception, for example).

Property: An existent reality that is universal, immutable, and can (or perhaps must) be *in* or *had* by other things more basic, such as a substance. Thus, a cow (a substance) can have the property of being brown. The brownness (property) is had by the cow (the substance).

Property dualism: A human being is one material substance that has both physical *and* mental properties, with the mental properties arising from the brain.

Property ontological reduction: One property (heat) is identified with another property (mean kinetic energy).

Proposition: A declarative sentence that is either true or false. Examples of propositions include: "The earth orbits the sun," "Greg is six feet tall," and "I lived in Canada when I was seven."

Propositional attitude: An attitude (such as hoping, fearing, wishing, regretting) toward a certain proposition. For example, "I hope that the test will be cancelled," or "I fear that the economy is slowing down," or "I regret that I didn't have a second piece of cake."

Propositional knowledge: Knowledge that a proposition is true.

Qualia: A quale (plural, qualia) is a specific sort of intrinsically characterized mental state, such as seeing red, having a sour taste, feeling a pain.

Self-presenting property: A property, such as *being appeared to redly* (that is, experiencing an appearance of the color red), that presents both its intentional object (e.g., a red apple) and itself (the redness) to the subject experiencing it.

Spirit: That faculty of the soul through which the person relates to God.

Structural supervenience: The view that mental properties are structural properties entirely constituted by the properties, relations, parts, and events at the subvenient level.

Substance: a particular, individual, continuant and basic, fundamental existent thing that is a unity of parts, properties, and capacities, and has causal powers.

Substance dualism: A human person has both a brain that is a physical thing with physical properties and a mind or soul that is a mental substance and has mental properties.

Supervenience: A relationship of dependence between properties such that one level of the properties correlates to conditions at a different level. For example, when water molecules come together, the property of wetness supervenes upon them. In mind/body discussions, some philosophers (such as certain property dualists) hold that mental events supervene upon (or emerge from) brain events.

Temporary disembodiment view: A person is (or has) an immaterial soul/spirit deeply unified with a body that can enter a temporary intermediate state of disembodiment at death, however unnatural and incomplete it may be, while awaiting a resurrection body in the final state.

Theoretical or explanatory reduction: One theory or law is reduced to another by biconditional bridge principles (for example, x has heat if and only if x has mean kinetic energy).

Thomistic substance dualism: The view that the (human) soul diffuses, informs (gives form to), unifies, animates, and makes human the body. The body is not a physical substance, but rather an ensouled physical structure such that if it loses the soul, it is no longer a human body in a strict, philosophical sense.

Thought: A mental content that can be expressed in an entire sentence and that only exists while it is being thought.

Token physicalism: Fundamentally, the claim that every token (that is, particular) mental event is identical to a particular physical event.

Traducianism: Both the body and soul are passed on to us by our parents.

Type-Identity physicalism: The view that mental properties/types are identical to physical properties/types.

Universalism: The view that God will eventually reconcile all things to himself, including all individuals, even if this means that God will continue to draw them to Himself in the afterlife.

IF YOU ENJOYED *THE SOUL*, YOU'LL LOVE THESE OTHER TITLES!

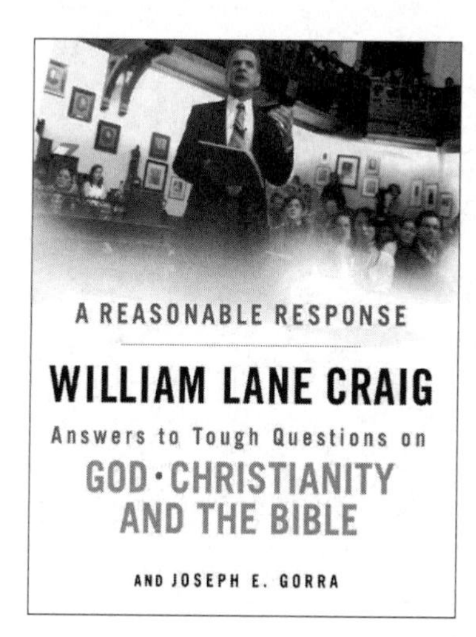

978-0-8024-0599-9

978-0-8024-1039-9

Also available as an ebook

www.MoodyPublishers.com